The Medici Effect

The
Medici Effect

**WHAT ELEPHANTS AND EPIDEMICS
CAN TEACH US ABOUT INNOVATION**

Frans Johansson

HARVARD BUSINESS SCHOOL PRESS

Boston, Massachusetts

Printed in the United States of America

10 09 08 07 06   5 4 3 2 1

ISBN: 1-4221-0282-3
ISBN13: 978-1-4221-0282-4

Library of Congress Cataloging-in-Publication Data forthcoming.

To my parents
For supporting every idea I've decided to pursue

Contents

Acknowledgments xi

Introduction 1

PART ONE

The Intersection

one

The Intersection—Your Best Chance to Innovate 11

Monkeys and Mind Readers

two

The Rise of Intersections 21

The Sounds of Shakira and the Emotions of Shrek

PART TWO

Creating the Medici Effect

three

Break Down the Barriers Between Fields 35

Sea Urchin Lollipops and Darwin's Finches

four

How to Make the Barriers Fall   45

Heathrow Tunnel and Restaurants Without Food

five

Randomly Combine Concepts   61

Card Games and Sky Rises

six

How to Find the Combinations   73

Meteorite Crashes and Code Breakers

seven

Ignite an Explosion of Ideas   89

Submarines and Tubular Bells

eight

How to Capture the Explosion   103

MacGyver and Boiling Potatoes

PART THREE

Making Intersectional Ideas Happen

nine

Execute Past Your Failures   119

Violence and School Curricula

ten

How to Succeed in the Face of Failure   127

Palm Pilots and Counterproductive Carrots

eleven

Break Out of Your Network   143

Ants and Truck Drivers

twelve

How to Leave the Network Behind   153

Penguins and Meditation

thirteen

Take Risks and Overcome Fear   161

Airplanes and Serial Entrepreneurs

fourteen

How to Adopt a Balanced View of Risk   171

Elephants and Epidemics

fifteen

Step into the Intersection . . .   183

and Create the Medici Effect

Notes   191

Index   201

About the Author   207

Acknowledgments

ALTHOUGH MY ENTIRE LIFE, in some ways, has pointed to this book, the idea that I should write it struck me one morning out of nowhere. I admit that I get ideas that way all the time, but only a fraction of them actually end up happening. This one did—and for that I am truly thankful to those who supported me during the process.

First and foremost, I wish to thank Teresa Amabile, a leading creativity researcher and my former professor at Harvard Business School. Teresa's early enthusiasm for the book gave me the inspiration I needed to get it off the ground. The first time we met to talk about this project she asked me if I knew just what I was getting myself into. Sure, I said. But, of course, I had no clue. I never realized just how much work it would be to pull together countless hours of interviews and thousands of pages of research. In the end it was my passion for this topic that made it possible—just as Teresa would have predicted.

While writing this book I got the opportunity to meet with some truly amazing people who graciously offered to share their experiences with me. I am deeply thankful to all of them. Although most of them appear in the book, some do not, and I want to recognize these people here. Thanks to Marcus Ahman, Susan Bird, Marguerite Bouvard, Martha Brass, Gregory Costikyan, Tobias Dahl, Edie Fraser, Bruno Giletti, Jim Head, Rosabeth Moss Kanter, Arcadia Kim, Paul Lawrence, Luke Visconti, and Michael Wheeler.

Many close friends took precious time to read the manuscript and share with me their thoughts and comments on how to improve it. Kristian Ribberström, Mark Tracy, and Tom Gates went way beyond the call of duty in this regard. Special thanks to Kristian Ribberström for letting us adapt his illustration for figure 7-1. I am also incredibly grateful for all the help, feedback, and support I got from Chantal Yang, Lisa Onaga, Roland Adolfsson, Martin Johansson, Raphael Brown, Elin Andersson, Chris Yeh, and Ahmed Alireza.

I was fortunate to work with a great group of people at HBS Press. Many thanks to Hollis Heimbouch, who not only saw the book's potential but also believed that I could pull it off. And endless thanks to my editors Jacque Murphy and Astrid Sandoval, who helped me to meet those expectations. Special thanks to Jacque who, with almost uncanny precision, pointed out the problems in my manuscript—and then gently suggested how I should correct them. Without her the text would still be in the Stone Age.

My deepest thanks to my aunt Lena, who lent me her summerhouse on an island off the coast of Sweden for a couple of stunning months in the fall. It was the perfect place to start writing this book. I am also deeply grateful to my sister, Sandra, who helped me parse each sentence in the manuscript for days and nights as the deadlines loomed. She could look me straight in the face and tell me that an entire chapter had to be revised twelve hours before deadline. And she would be right. Thanks also to my cousin Christian for being who he is—the closest of friends.

To my parents, my deepest thanks. Throughout life they have been my greatest source of inspiration, no matter what type of enterprise I pursued. For this I owe them more than they will ever know. They are the truest example of a glorious life at the Intersection.

Finally, I want to thank my fiancée, Sweet Joy. She helped me see the difference between what I wanted to say and what I was actually saying—and she helped me close the gap. We started dating the same week I got the idea for this book, and she has tolerated and supported all the idiosyncrasies of a very focused writer with an almost incomprehensible enthusiasm. For that, and a great many other things, I am eternally grateful.

The Medici Effect

Introduction

PETER'S CAFÉ sits on a hillside in Horta, a port city on one of the Azores islands in the middle of the Atlantic Ocean. By the time you reach the docks in the harbor, you can tell that this place is special. Bright, colorful paintings of sailboats and flags line the piers—hundreds and hundreds of them, drawn by visiting captains and crew members from every corner of the globe. Horta is the one place between the Americas and Europe where world-traveling sailors stop to take a break. Some are heading toward Fiji, others to Spain. Some are on their second tour around the world; others are simply resting before the last leg to Brazil. They come from different backgrounds and cultures. And all of them converge upon the rustic-looking Peter's Café. Here they can pick up year-old letters from other world travelers or just sit and talk over a beer or a glass of Madeira.

When I saw this place for the first time, I realized that the serene environment of the café actually concealed a chaotic universe. The café was filled with ideas and viewpoints from all corners of the world, and these ideas were intermingling and colliding with each other.

"Get this, they don't use hooks when fishing for marlin in Cuba," one visitor says.

"So what do they use?" another asks.

"Rags. The lure is covered in rags. When the fish strikes the rag, it wraps around the fish bill and won't let go because of the friction. The fish don't get hurt and can be released, no problem."

"That's pretty neat. Maybe we could use something like that. . . ."

The people here participate in what seems like an almost random combination of ideas. One conversation leads into another, and it is difficult to guess what idea will come up next. Peter's Café is a nexus point in the world, one of the most extreme I have ever seen.

There is another place just like Peter's Café, but it is not in the Azores. It is in our minds. It is a place where different cultures, domains, and disciplines stream together toward a single point. They connect, allowing for established concepts to clash and combine, ultimately forming a multitude of new, groundbreaking ideas. This place, where the different fields meet, is what I call the Intersection. And the explosion of remarkable innovations that you find there is what I call the Medici Effect. This book is about how to create it.

Creating the Medici Effect

THE IDEA BEHIND THIS BOOK is simple: When you step into an intersection of fields, disciplines, or cultures, you can combine existing concepts into a large number of extraordinary new ideas. The name I have given this phenomenon, the Medici Effect, comes from a remarkable burst of creativity in fifteenth-century Italy.

The Medicis were a banking family in Florence who funded creators from a wide range of disciplines. Thanks to this family and a few others like it, sculptors, scientists, poets, philosophers, financiers, painters, and architects converged upon the city of Florence. There they found each other, learned from one another, and broke down barriers between disciplines and cultures. Together they forged a new world based on new ideas—what became known as the Renaissance. As a result, the city became the epicenter of a creative explosion, one of the most innovative eras in history. The effects of the Medici family can be felt even to this day.

We, too, can create the Medici Effect. We can ignite this explosion of extraordinary ideas and take advantage of it as individuals, as teams, and as organizations. We can do it by bringing together different disciplines and cultures and searching for the places where they connect. *The Medici Effect* will show you how to find such intersectional ideas and make them happen. This book is not about the Renaissance era, nor is it about the Medici family. Rather, it is about those elements that made that era possible. It is about what happens when you step into an intersection of different disciplines and cultures, and bring the ideas you find there to life.

Surprising Insight

MICK PEARCE, an architect with an interest in ecology, accepted an intriguing challenge from Old Mutual, an insurance and real estate conglomerate: Build an attractive, functioning office building that uses no air conditioning. Oh, and do it in Harare, the capital of Zimbabwe.[1]

This may, on the face of it, seem ridiculous. After all, it can get pretty hot in Harare. But Pearce, born in Zimbabwe, schooled in South Africa, and trained as an architect in London, was up for the challenge. And he achieved it by basing his architectural designs on how *termites* cool their towerlike mounds of mud and dirt. What's the connection?

Termites must keep the internal temperature in their mounds at a constant 87 degrees in order to grow an essential fungus. Not an easy job since temperatures on the African plains can range from over 100 degrees during the day to below 40 at night. Still, the insects manage it by ingeniously directing breezes at the base of the mound into chambers with cool, wet mud and then redirecting this cooled air to the peak. By constantly building new vents and closing old ones, they can regulate the temperature very precisely.

Pearce's interests clearly extend beyond architecture. He also has a passion for understanding natural ecosystems, and suddenly those two fields intersected. Pearce teamed up with engineer Ove Arup to bring this combination of concepts to fruition. The office complex, called Eastgate, opened in 1996 and is the largest commercial/retail complex in Zimbabwe. It maintains a steady temperature of 73 to 77 degrees and uses less than 10 percent of the energy consumed by other buildings its size. In fact, Old Mutual saved $3.5 million immediately because they did not have to install an air-conditioning plant. Eastgate ultimately became a reference point for architects—articles and books have been written about it, and awards have been given. Mick Pearce is known as a groundbreaking innovator for launching a new field of architectural design—one that "copies the processes of nature."[2]

How did Pearce come up with such an innovative design? Was it luck? Maybe; luck is part of everything we do. The more intriguing question is, what did Pearce do to affect his chances of accomplishing this breakthrough? Did he, in effect, make his own luck? The answer is yes, and the reasons why lie at the heart of this book's message. Pearce had stepped into the Intersection, a place where he could combine architectural designs with processes in nature. It was his willingness to explore these combinations that made it more likely for him to successfully break new ground. The Intersection is certainly not the only place to uncover new ideas, but I'll argue that it is the best place to generate and realize extraordinary ones.

A Place for Everyone

MICK PEARCE is one example of a person who found the Intersection and made successful discoveries there. From this example one might get the impression that the Intersection is a place only for designers and artists. It's easy to associate creativity with art, but creativity includes new ideas in every field, from science and business to law and politics.

Consider, for instance, the seeming antithesis of the idealistic artist, George Soros, one of the most respected investors of our time. He is perhaps best known as the man who broke the Bank of England in 1992. Soros made a profit of over $1 billion in *one afternoon* by betting that the pound sterling was overvalued. Although he has also had some stinging losses, Soros's track record as an investor is astonishing, having generated billions for his fund.

Perhaps his most important legacy, however, will not be the money he accumulated for his limited partner but his ideas about democracy, his philosophy concerning capitalism, and his approach to philanthropy. Soros pulled together ideas from the fields of finance and philosophy to create an innovative philanthropic strategy. That strategy, which was unprecedented in its audacity, focused on transforming nations into societies that are based on the recognition that nobody has a monopoly on the truth—what he calls "Open Societies." Michael Kaufman writes in *Soros: The Life and Times of a Messianic Billionaire* about the exploratory journey Soros took to understand the world this way: "In the process, he digressively took up dozens of themes, among them the limits of knowledge, the development of modern art, the flaws of classical economics, the value of fallibility, and even the prospects of fundamental reforms in the Soviet Union."[3]

George Soros found the Intersection. He found a way to connect completely separate fields and he managed to do so in a meaningful way. Just like Mick Pearce.

Connections Everywhere

THIS MAY ALL SOUND somewhat improbable. Can great innovative breakthroughs, those that can create a Medici Effect, be explained by the intersection of disciplines and cultures? And if so, is it possible to understand the nature of this intersection and to harness its power? The answer is yes, on both counts. In writing *The Medici Effect* I have three objectives:

1. The first is to explain what, exactly, the Intersection is and why we can expect to see a lot more of it in the future. You will see how three critical forces are working together to increase the number of intersections around the world.

2. The second is to explain why stepping into the Intersection creates the Medici Effect. You will see why it is such a vibrant place for creativity and how we can use intersections to generate remarkable, surprising, and groundbreaking ideas.

3. Finally, the third objective is to outline the unique challenges we face when executing intersectional ideas and how we can overcome those challenges. You will see how execution at the Intersection is different from within established fields, and you will learn how to prepare for those differences.

In order to fulfill these three objectives, I have relied on the work of leading researchers in creativity and innovation, such as Dean Keith Simonton, Clayton Christensen, Teresa Amabile, and Robert Sutton, and on a range of psychologists, economists, and sociologists. My most interesting discoveries and conclusions, however, have come from numerous conversations and interviews with people who live and operate at the Intersection. The stories of how they found their way to the Intersection, and how they created the Medici Effect, contain enough surprises and valuable insights to easily fill two or three books.

You will, for instance, meet a mathematician from Seattle who stepped into the intersection of games and collectibles to create one of the world's fastest-spreading recreational activities. You will learn how he did it and why those lessons hold true for anyone at the Intersection. You will read about an entrepreneur who steps into the Intersection every time he starts a new company. His story will show you how we can find courage at the brink of uncertainty. You will encounter a physician who made the connection between violence prevention and health care. No one else understood the link at the time, and her struggle to bring her ideas to life demonstrates the challenges anyone will face at the Intersection. During this journey you will also meet a woman who hiked through a snake-infested prisoner island off the coast of Colombia while gathering lava rocks for her research. You will read about a chef who surprised the world with his food concoctions at the age of twenty-four and learn about a team of researchers who discovered how to read the mind of a monkey.

These individuals and their remarkable acts of innovation help us understand the power of the Intersection. They have all managed to connect fields we thought were unrelated. When they did, they generated ideas that changed them, their organizations, and, ultimately, a part of our world. From these examples, we can learn how to do the same. Their stories answer the central questions this book poses: How do we create an explosion of extraordinary ideas, and how do we make those ideas happen? The answers may surprise you.

PART ONE

The Intersection

CHAPTER one

The Intersection—
Your Best Chance to Innovate

MONKEYS AND MIND READERS

I N THE SPRING OF 2002, a team of researchers at Brown University in Providence, Rhode Island, conducted a remarkable experiment.[1] The experiment went something like this: A rhesus monkey is trained to play a computer game. The point of the game is to use a yellow cursor to chase down a red dot that moves randomly across the screen like an erratic hockey puck. The game looks and feels like something designed for a child except for one noticeable difference. The monkey doesn't use a mouse or a joystick to play this game. Rather, the monkey moves the cursor with its mind. It controls where the cursor goes—mentally.[2]

When these results were published in the prestigious science journal *Nature*, they became what was likely the most reported Brown University science story ever.[3] The day the press release circulated over the wires, Mijail Serruya, the graduate student behind the experiments, was flooded with calls from every corner of the globe. "I'm on the way to the bathroom to brush my teeth, half asleep," Serruya recalls, "and it's 'Hello. This is the BBC.'" Reporters wanted to know everything from whether people could use the technology for military contraptions to whether it could help a "couch potato" get off his butt.

This story is especially compelling not just because of what the team of scientists discovered, but also because it was a result of a deliberate effort to find an intersection of disciplines. The group behind this particular breakthrough consisted of mathematicians, medical doctors, neuroscientists, and computer scientists, all playing crucial roles in understanding how the brain works. The team was firmly planted at the Intersection—and they struck gold because of it.

This was no accident. Professor Leon Cooper, who pioneered the brain science research efforts at Brown University, made a special point of bringing together a wide range of disciplines to understand the human mind.[4] Cooper himself has a broad set of interests. When he received the Nobel Prize for his work in solid-state physics, almost three decades before the "mind-reading" experiment, he had already switched fields once. He had moved into brain science and founded, among other things, Nestor, Inc., one of the very first neural networking companies in the United States.[5] Cooper had witnessed the awesome benefits of bringing different fields together and made it an essential part of the Brain Science Program's strategy. "Brain research is different [from] pure physics research. The nature of the beast is that you have to put together a different kind of team," Cooper told me one afternoon. "Our interdisciplinary approach sets us apart and gives us a chance to lead new discovery in this area." The mind-reading experiment is an excellent example of what he was talking about.[6]

The team had in this case managed to "eavesdrop" on the part of the brain that plans motion. Tiny implanted electrodes read signals from the monkey's brain cells, which a computer deciphered through advanced statistical techniques. What was once a lot of incomprehensible data from the brain could now be translated into what the monkey was thinking. As a result, the team could turn thoughts into action in real time. This incredible breakthrough was a result of different people from different fields coming together to find a place for their ideas to meet, collide, and build on each other.

The implications of the discovery are enormous. "This implant is potentially one that is very suitable for humans," says Mijail Serruya. "It shows enough promise that we think it could ultimately be hooked up via a computer to a paralyzed patient to restore that individual's interaction with the environment." Looking into the future, Serruya says, a prosthetic arm that moves by thoughts alone is no longer just a sci-fi dream.[7]

Today the Brain Science Program, now headed by John Donoghue, consists of researchers in the cognitive sciences, neuroscience, computer science, biology, medicine, psychology, psychiatry, physics, and mathematics. Both Donoghue and Cooper believe it is critical to step into the intersection of these diverse fields to achieve the breakthrough ideas that will push discoveries forward. "For instance, unexpectedly bumping into a statistician in the hallway one afternoon can lead to a discussion that solves a particular problem I have been struggling with," Donoghue explains. The researchers are not quite sure when something interesting will happen, but if they keep talking, they know that something eventually will.[8]

The same approach that led this team of scientists to groundbreaking discoveries is, at its root, the same approach that led to the unique architectural designs of Mick Pearce and the investment/philanthropic strategies of George Soros. But why does such an approach have a better chance of radically changing the world than any other? Before we can answer that question, we must first understand something about the nature of creative ideas and the process of innovation.

Creative Ideas and Innovation

WHY, EXACTLY, do we call the experiments made by the team at the Brain Science Program innovative? The fact that most people get their socks knocked off when they see the rhesus monkey play the game is not enough. We can be wowed by any number of things, from the size of the world's largest pumpkin to a 5 P.M. Los Angeles traffic jam—but that doesn't mean they're innovative.

Here's why: The mind-reading experiment was creative because it was *new* and *valuable*, and it was innovative because the creative idea had become *realized*. This definition of creativity and innovation aligns most closely with that posed by leading Harvard Business School creativity researcher Teresa Amabile.[9] Although the definition may seem obvious, it is worth spending some time to examine it more closely.

Creative Ideas Are New

The team behind the experiments had accomplished something unique, something no one had done before—clearly a key characteristic of a creative idea. If you duplicate a painting by Monet you have not done something creative, and if you set up a bookshop Web site that operates exactly like Amazon.com, you have copied a business model, not innovated.

This criterion seems obvious, but it can be deceptive in its simplicity. What if an idea is new to the creator, but not to others? Unfortunately, it would be hard to consider such an idea innovative. Imagine, for instance, if someone claimed to have discovered the double-helix structure of DNA. No one would pay any attention. Watson and Crick did that more than fifty years ago. But what if the situation is the reverse? What if the idea is old to the creator, but new to others? The creator could, for instance, tell an old story in a new rendition, or use a screw cap in a new fashion (as Thomas Edison did when he and his team developed the fixture for the light bulb). In such a case society will agree that the product is indeed creative. In fact, most creative activity happens in this way.[10]

Creative Ideas Are Valuable

Interestingly, to be considered creative, it is not enough that an idea is new. To say that $4 + 4 = 35,372$ is definitely original, but it hardly qualifies as creative.[11] For an original idea to be creative, it must also have some measure of relevance; it must be valuable. Saying that 4 + 4 = 44 while keeping a straight face (as Chris Rock did in his movie *Head of State*) could fulfill such a requirement, since some people may find it amusing. This, then, explains why the experiment made by the brain science team was creative. It was new and valuable to a fairly large number of people, as clearly indicated by the publication of the research in *Nature* and the media onslaught that followed.

Innovative Ideas Are Realized

The reason we call the team's experiment innovative is that they made it happen, and others are now using the discoveries to further their own research. Innovations must not only be valuable, they must also be put to use by others in society. Simply imagining the most amazing invention ever does not qualify one as an innovative person. If an idea exists solely in someone's head, it cannot yet be considered innovative. It has to be "sold" to others in the world, whether those people are peers who review scientific evidence, customers who buy new products, or readers of articles or books.

In some ways this generally accepted definition of creativity and innovation is a bit disconcerting. Usually we think of individuals as creative, but creativity really occurs when people act in concert with the surrounding environment, and within society.[12] Ultimately society decides whether an idea is both new and valuable. In the words of psychologist and leading creativity researcher Mihaly Csikszentmihalyi, "There is no way to know whether a thought is new except with reference to some standards, and there is no way to tell whether it is valuable until it passes social evaluation."[13] Thus, it is impossible to determine if a person's products are innovative if they have never been seen, used, or evaluated.

Having built some boundaries around the world we will explore here, let's drill back down. This book argues that the Intersection is the best place to generate an explosion of new breakthrough ideas—what I call the Medici Effect. But what, exactly, *is* the Intersection?

The Intersection: Where Different Fields Meet

WHEN WE SAY that the Brain Science Program sits at the intersection of mathematics and medicine, of computer science and neurophysiology, what we are really saying is that *the people in the program have managed to connect these fields, and through these connections they have come up with new creative insights*. Individuals, teams, or organizations step into the Intersection by associating concepts from one field with concepts in another. The Intersection, then, becomes a virtual Peter's Café, a place for wildly different ideas to bump into and build upon each other.

The term *field* is used in this book to describe disciplines, cultures, and domains in which one can specialize through education, work, hobbies, traditions, or other life experiences. Fields can, for instance, include mystery writing, painting, Chinese business customs, molecular biology, and the enterprise software industry. They encompass areas as diverse as sport fishing, cable television, Hispanic-American culture, equity analysis, object-oriented programming, poetry, carpeting, and movie editing. Fields can, in turn, be divided into a subset of more narrowly defined fields. For instance, you can talk about the field of cooking generally, but you can also talk about the specialties of Swedish and Thai cuisine. Ultimately, in order for an area to be called a field, a person should conceivably be able to spend a lifetime involved with it.

Fields consist of *concepts* such as knowledge and practices. Changing a tire can be called a concept. So can the item *tire*, in and of itself. These two concepts are both included in a field called mechanics. In order to understand a field, one has to understand at least some of its concepts. The more concepts one understands within a field, the more expertise one has built within that field.

The key difference between a field and an intersection of fields lies in how concepts within them are combined. If you operate within a field, you primarily are able to combine concepts within that particular field, generating ideas that evolve along a particular direction—what I call *directional* ideas. When you step into the Intersection, you can combine concepts between multiple fields, generating ideas that leap in new directions—what I call *intersectional* ideas. The difference between these two types of ideas is significant.

Intersectional Ideas Will Make You Do a Double Take

THE EVOLUTIONARY BIOLOGIST Richard Dawkins is well known in his field. In 1976 he published *The Selfish Gene*, a book that pushed evolutionary theory a big step forward. Dawkins suggested that evolution did not occur between species or even between organisms, but between genes—and that these genes were "selfish." This theory was a notable contribution to his field and earned Dawkins significant acclaim.[14]

It is therefore rather curious to note that Dawkins's arguably most widespread contribution to society was a very different type of idea, one that originated from a single, fairly off-topic chapter in his book. In it Dawkins connected the field of genetic evolution with that of cultural evolution—and made the connection explicit. He suggested that ideas, which are the building blocks of our culture, evolve and propagate just like genes. He called these building blocks *memes* and wrote:

> *Examples of memes are tunes, ideas, catch-phrases, clothes fashions, ways of making pots or of building arches. Just as genes propagate themselves in the gene pool by leaping from body to body via sperm and eggs, so do memes propagate themselves in the meme pool by leaping from brain to brain via a process which, in the broad sense, can be called imitation.*[15]

Most people I know did a double take while reading this chapter by Dawkins. What an incredible notion! Ideas, or memes, compete, in a real sense, for space in our minds. Some memes persist and transform, others die out; the process is similar to that of genetic evolution. Not only did this notion seem to make intuitive sense, it was *cool*. And it came from an intersection.

Dawkins's first idea about the selfish gene was directional; his second idea about memes was intersectional. The first idea pushed an established field farther along an established direction; the second idea came out of nowhere, ultimately launching a field of its own—memetics.[16]

The concept of the meme took off almost immediately and has today become a way for marketers, sociologists, and historians to explain, predict, and affect cultural phenomena.[17] For example, in his best-selling book *The Tipping Point*, Malcolm Gladwell examines how the Hush Puppy was transformed from a dowdy shoe with stagnant sales to a hot fashion accessory in just a couple of years through a process best understood as an epidemic of an idea virus. Today many marketing strategies are based on the notion that ideas and fads act as a virus while spreading through a population of minds.[18] These strategies are a direct result of Dawkins's intersectional insight during the mid-seventies. Intersectional innovations, like the meme, are often more powerful and widespread than directional ones, but it is important to note that both types are needed for long-term success.[19] Why?

Two Types of Ideas

THE MAJOR DIFFERENCE between a directional idea and an intersectional one is that we know where we are going with the former. The idea has a *direction*. Directional innovation improves a product in fairly predictable steps, along a well-defined dimension. Examples of directional innovation are all around us because they represent the majority of all innovations. Consider, for instance, a company that improves efficiency by streamlining and refining an existing process, a scientist who defines a particular phenomenon to its sixth decimal (after knowing its fifth), or a successful policy program from one city that is tailored to fit into another. The goal is to evolve an established idea by using refinements and adjustments. The rewards for doing so are reasonably predictable and attained relatively quickly.

People and organizations do this all the time through increasing levels of expertise and specialization. It is absolutely necessary if one does not wish to squander the value of an idea. Even an intersectional idea will, once it has become established, develop and evolve along a specific direction. When Stephen Covey, author of the widely popular self-empowerment book *The Seven Habits of Highly Effective People*, released *The Seven Habits of Highly Effective Families*, he most likely did not intend to introduce a radically different idea, but to present the original idea with adjustments (and continue to reap rewards from it). The same holds true for companies that refine their products to new market segments, for researchers who delve deeper into an established field, and so on.

Intersectional innovations, on the other hand, change the world in leaps along new directions. They usually pave the way for a new field and therefore make it possible for the people who originated them to become the leaders in the fields they created. Intersectional innovations also do not require as much expertise as directional innovation and can therefore be executed by the people you least suspect. Although intersectional innovations are radical, they can work in both large and small ways. They can involve the design of a large department store or the topic of a novella; they can include a special-effects technique or the product development for a multinational corporation. In summary, intersectional innovations share the following characteristics:

> ➤ They are surprising and fascinating.
> ➤ They take leaps in new directions.
> ➤ They open up entirely new fields.
> ⊘ They provide a space for a person, team, or company to call its own.
> ➤ They generate followers, which means the creators can become leaders.

> They provide a source of directional innovation for years or decades to come.

> They can affect the world in unprecedented ways.

The Intersection Is Your Best Chance to Innovate

FOR MOST OF US, the best chance to innovate lies at the Intersection. Not only do we have a greater chance of finding remarkable idea combinations there, we will also find many more of them. To be specific, stepping into the Intersection does *not* mean simply combining two different concepts into a new idea. These types of combinations are part of both directional and intersectional innovation. Instead, the Intersection represents a place that drastically *increases* the chances for *unusual* combinations to occur.

Imagine that you are a health care worker caring for paralyzed patients. If you wish to develop new treatment strategies from within your field, you have to understand that field thoroughly. It is critical that you master most concepts within your field to find new ideas that work. In addition, since it is easy to predict where the field is heading, you will have a lot of competition at every turn.

Now imagine that you reach out and connect your experience with that of neuroscience. Suddenly there will be many new options and ideas for you to explore. Neurological concepts you had no idea even existed can potentially be combined with existing treatment strategies to generate breakthrough intersectional ideas. By stepping into the Intersection you will, in other words, have unleashed an explosion of fresh, intriguing idea combinations.

This explosion of remarkable ideas is what happened in Florence during the Renaissance, and it suggests something very important. If we can just reach an intersection of disciplines or cultures, we will have a greater chance of innovating, simply because there are so many unusual ideas to go around. And as the following chapter will show, there has never been a better time to do it than now.

The Rise of Intersections

THE SOUNDS OF SHAKIRA AND
THE EMOTIONS OF SHREK

THE STORY about the brain science team and their rhesus monkey is a story of our times. It reflects an increasingly interconnected world where concepts that appear to be unrelated actually are related. It reflects a rise of intersections. This type of story should not surprise us—we will see more like it. Lots more, everywhere.

It is certainly not the first time we have seen such a convergence among fields. Leonardo da Vinci is the illustrious standard-bearer of the Renaissance, when artists, scientists, and merchants stepped into the Intersection together and created one of Europe's most creative explosions of art, culture, and science. But the centuries that followed saw a growing specialization of knowledge. Disciplines became more fragmented as we broke the world into smaller and more specialized pieces. Today, however, that fragmentation is reversing and the effects can be seen in fields everywhere. Tom Friedman, foreign affairs columnist at the *New York Times*, comments on the growing connections in today's world in his book *The Lexus and the Olive Tree*: "Today, more than ever, the traditional boundaries between politics, culture, technology, finance, national security and ecology are disappearing."[1]

There are three distinct forces behind the rise of intersections, and at this moment, perhaps for the first time, they are all working together. They are not the only reasons that intersectional innovations happen, but they explain why we are seeing more of them than ever.

Force 1: The Movement of People

IN 1809 a mixed-blood Cherokee Indian named Sequoyah learned to sign his name on his silversmith work. That was his introduction to the written language. A few years later, while serving in the U.S. Army during the Creek War, he saw American soldiers write letters, read orders, and record historical events of the war. Sequoyah realized that his fellow men in Cherokee Nation could derive spectacular benefits from a written language. Sequoyah, whose mother was a member of the Paint Clan and whose father was a Virginia fur trader, spent the next twelve years developing a written Cherokee language. When he was done he had constructed a syllabary that consisted of eighty-five characters representing each syllable in the Cherokee language. The syllabary was so easy to learn that within weeks thousands of Cherokees could read, and it gave Cherokee Nation the ability to create the first Native American newspaper, *The Cherokee Phoenix*. Sequoyah is the only person in the world known to have created an entire written language on his own and is considered a genius to this day.[2]

Sequoyah got the idea for creating a written language after spending time in a culture very different from his own. This is one method of finding intersections (one that will be explored in greater detail in the next chapter), but it is also an example of the force of globalization. That force, as defined by the movement of people between cultures and countries, is regaining a strength it has not had for more than a hundred years.

The movement of people is on the rise everywhere, for several reasons. The widespread effects of democracy and capitalism, with its lower trade barriers and open nation borders, have led to an increase in jobs and education for foreigners in most countries around the world. In addition, the flow of refugees and political asylum seekers remains quite strong. Other factors even suggest an accelerated rate of movement. Consider, for instance, that virtually all industrialized countries face a population shortfall, endangering the social security systems in those countries. The rapidly aging population and dwindling birth rates can arguably be compensated only through increased levels of immigration.[3] Clearly, though, the movement of people is on the rise and it can be detected in the census numbers from countries around the world.

In the United States, for instance, the percentage of foreign-borns has risen to levels not seen since the 1930s. According to the 2000 census, 11.1 percent of the U.S. population is foreign-born—an almost 60 percent increase since 1990. This upward trend started in 1970, when less than 5 percent of the population was foreign-born. The trend is not exclusive to the United States—it is happening everywhere. Just between 1994 and 1999, the foreign-born population grew between 5 and 17 percent in countries such as Korea, Denmark, Spain, Australia, Italy, and Canada.[4] According to management guru Peter Drucker, "The mass migrations of the nineteenth century were either into empty, unsettled spaces (such as the United States, Canada, Australia, Brazil), or from farm to city in the same country. By contrast, immigration in the twenty-first century is by foreigners—in nationality, language, culture and religion—who move into settled countries." Drucker sees little reason to expect this long-term trend to reverse itself.[5]

This force will lead to a plethora of cultural intersections and a host of groundbreaking ideas for those bold enough to explore them. Cross-cultural ideas will be more easily introduced to a more diverse audience. This holds particularly true for businesses and the arts. When the Latin American artist Shakira made her U.S. debut with the album *Laundry Service*, she shot to the top of the charts. Her music had been unusual even in her home country of Colombia. Her father is Lebanese, and her songs combined Arabic and Latin sounds into "a distinctive blend of pop and rock unlike anything being done by Colombian singers at the time."[6] She managed to take this innovative music and intersect it with American tunes. *Newsweek* wrote:

> *young stars like . . . the Colombian rocker Shakira break down the divisions by mixing a variety of pop styles, Latin and Anglo. "We are made of fusion," says Shakira, 22. "It's what determines our identity: the way in one mouthful we take rice,* plátanos, *meat." Her own music combines Alanis Morissette, reggae and Mexican mariachi sounds.*[7]

These trends—the blending and mixing of cultures—are becoming more evident every year in fields such as cinema, literature, music, and art. Businesses, too, will increasingly be able to innovate in different regions of the world. They can arbitrage ideas between different cultures by understanding how those cultures connect. This holds true not just for major corporations but also for your neighborhood store.

One day, for instance, I was walking along Fifth Avenue in Brooklyn, New York, when I noticed a store named Kimera, after the Greek monster that was a hybrid of a lion, goat, and snake. It turned out to be a clothing store, one with a very distinctive style. One shirt, for example, looked like a blend of a kimono and a standard Western-style blouse. Other garments were similar hybrids.

The founder, a woman named Yvonne Chu, told me she drew inspiration from her experiences growing up in New York with her Chinese parents as well as from her world travels. People love her unique designs and the mix of cultures so apparent in them.[8] "This shirt," she said, pulling out a purplish-blue shirt with a Mandarin collar, brocade bottom, and front ties, "people just went crazy for it." Kimera is a sign of the times. The movement of people across countries and cultures is creating more intersections than ever.

Force 2: The Convergence of Science

S MITHSONIAN MAGAZINE ran a story that caught my attention. Biotechnicians had inserted a silk-producing gene from a golden orb weaver spider into a herd of goats. The point was for the goats to produce milk that contained the essence of spider webs, a material with amazing strength. The researchers, in turn, could use the milk to "spin" threads with silk-like qualities. Amazing, but true.

> *Pound for pound, the gossamer silk threads created by orb weaver spiders are five times stronger than steel. One day, [CEO Jeffrey] Turner believes, spider silk might be found in everything from air bags, fishing line and non-tear sports jerseys to ophthalmic sutures and artificial tendons.*[9]

The company behind this innovation, Nexia, had completed the largest IPO in Canadian life sciences history and was already stepping up production. This story harks back to the mind-reading experiment at Brown University because both of them are indicators of what is happening within the sciences. The nature of scientific discovery is changing. The disciplines that once were so separate are coming together again.

Consider this: How many times can you discover a continent? In the case of America, we know that it's been done at least three times by three representatives from three different civilizations. The Native Americans crossed the land bridge at the Bering Strait in three consecutive waves some twenty thousand years ago; the Vikings went from Iceland via Greenland to Newfoundland about a thousand years ago; and finally, Christopher Columbus accomplished the same feat, albeit at more southern latitudes, around five hundred years ago. Today, however, such a discovery is impossible. It's been done. The big discoveries within traditional geography have been made and are well documented. The same, of course, holds true of human anatomy. What if this is true for the rest of the sciences?

In field after field, we are finding that our basic understanding of the world is, if not 100 percent accurate, at least good enough. Consider chemistry. The number of chemical variations may be impossibly large, but the principles that govern such variations are clearly limited and were mostly explained by Linus Pauling as early as the 1930s, an accomplishment that earned him one of his two Nobel Prizes (the other was for peace). In biology virtually every discovery, including the double helix, has reinforced and refined Darwin's theory of evolution, not questioned it. We have spent a lot of time segmenting the world, trying to understand its individual components, and we have done a good job at it. In short, science works, and it works well. However, just as there are a limited number of times that we can discover a continent or a section of human anatomy, we can discover the law of evolution, or a supernova, or thermodynamics, only once.[10]

This does *not* mean, however, that science has played out its role. On the contrary, science is becoming increasingly critical to all of our lives. There are more questions to explore than ever before, but a great many of the discoveries will be of a different nature than in the past. Instead of helping us understand the individual pieces of the world, they will help us understand how those pieces interact. So, for instance, you will find engineers collaborating with biologists to understand the toughness of the conch shell and applying it to everything from tank armor to auto bodies.[11] Or you will see oceanographers, meteorologists, geologists, physicists, chemists, and biologists collaborating to understand the effects of global warming. New discoveries, world-changing discoveries, will come from the intersections of disciplines, not from within them.

Scientists are increasingly recognizing this trend. I spoke with Alan Leshner—CEO of the American Association for the Advancement of Science (AAAS) and arguably one of the most influential and well-connected people in the scientific community—about the rise of intersections. AAAS is the world's largest science organization, and more than a million people around the world read its journal, *Science*, every week. I asked him what the future holds for scientific discoveries within disciplines.[12]

"Disciplinary science has died," he shot back. "It's gone." Leshner sees more and more evidence for such a conclusion. "Most major advancements involve multiple disciplines," he explained. "It is rarer and rarer to see single-author papers. And often the multiple authors are from different disciplines." This shift can also be seen in our colleges, where students today have many more hyphens in their majors than in the past. For instance, we now have college graduates in applied mathematics–physics, biology-chemistry, geology-chemistry, and economics-psychology. In addition, different departments are coming together to explore specific issues relating to the environment, bioengineering, sustainable development, and neuroscience, among many others.

Those scientists who understand the force of convergence are increasingly teaming up across disciplines. In perhaps no case has this happened with more success than at the Santa Fe Institute (SFI) in New Mexico. A man named George Cowan founded the SFI in 1984. He is a no-nonsense gentleman who speaks slowly, but with sharpness and wit in every sentence.[13] Whether the topic is art, business, or policy, he talks like he believes that science and mathematics are connected to everything—and that the Santa Fe Institute is, in a way, set up to find those connections.

Cowan believes firmly in the power of the Intersection, perhaps because he has seen its power many times. He worked as an administrator with top-notch scientists from every conceivable discipline during the Manhattan Project (for atomic bomb research). Since then he's served as associate director for research at Los Alamos National Laboratory while simultaneously leading a bank. It was during his time as a science advisor for the White House that he came up with the idea for the Santa Fe Institute. Cowan found himself struggling to bring scientists and politicians together. "I turned to one of the people that talked politics," Cowan says, "and asked, 'How do you get this kind of interaction to work?' 'Well, you have to learn their agenda,' he told me. 'How do you that?' I asked. 'You need to get scientists to think about things other than their specialty,' he answered."

Cowan created the Santa Fe Institute soon after that conversation. As the institute's charter mentions, it was formed to devote "itself to the creation of a new kind of scientific research community pursuing emerging syntheses in science." The SFI has been remarkably successful in this mission, and the research that emerges from it is as baffling as it is promising.

Biologists, for instance, can be found working together with economists and stock market analysts to generate new ideas about how markets behave. "The models we use to explain the evolution of financial strategies are mathematically similar to the equations biologists use to understand populations of predator-prey systems, competing systems, and symbiotic systems," says renowned investment manager Robert Hagstrom, vice president and executive director of Legg Mason Focus Capital.[14] Another well-known area of research is the small-worlds phenomenon, where people try to understand the world through the links that build it. These researchers see commonalities between the way body cells are structured, Web pages are linked, societies are shaped (like the famous theory of six degrees of separation), and even how terrorist cells interact.

Today the SFI is a private, independent research institution that allows researchers from the physical, biological, computational, and social sciences to collaborate. The Santa Fe Institute is another sign of the times. It is an institution that has grown out of the fact that science is reaching an inflection point—a time of convergence.

Force 3: The Leap of Computation

IN 2001 practitioners in the field of traditional 2D animation—what we think of as cartoons—realized their worst nightmares had become real. The nightmares came in the shape of a big green ogre named Shrek and a blue monster named Sulley. They were the main characters of two computer-animated 3D movies, both of which won great critical acclaim and stomped all over the competition at the box office. The companies behind the two movies, Dreamworks and Pixar, had brought 3D computer-animated movies into the big time. Although the technology had been around for over a decade, Steve Jobs's company Pixar took it to a whole new level. Pixar started out as a small animation shop, but after a couple of hit movies, including *Toy Story* and *A Bug's Life*, people begun taking the company seriously.

After the success of *Shrek* and *Monsters, Inc.* it did not take long for a debate to erupt among traditional animators. Was this the end? Would the hand-drawn creative artist go the way of the eight-track tape? No, some would say. It was just that the computer-animated movies that particular year had better stories and more interesting personalities than the traditional movies. There is definitely some truth to that. Both movies were simultaneously hilarious and sophisticated, enabling them to enthrall children and adults alike. The dialogue was witty, the emotional touches were striking (the creatures' eye movements were especially captivating), and the story compelling. So maybe the difference is not the computers; maybe it's the story and the way that story is told. But what if computers helped with the development of the story? Consider what Steve Jobs said in his first annual report for Pixar after it had gone public in 1996:

> In the new world of computer animation the opportunities for innovation are immense. . . . Traditional cell animators must spend a great deal of time drawing, because every one of the over 100,000 frames in a typical feature-length animated film (24 frames per second × 75 minutes) must be drawn by hand. In Pixar's computer animation, all the drawing is done by computers; hundreds and hundreds of very fast computers. This process results in . . . important differences from traditional cell animation. First, it frees our animators from drawing so that they can concentrate on acting, breathing life into their characters as they move. This allows Pixar to hire animators who may or may not excel at drawing, but are brilliant actors. Our animators even take acting lessons.[15]

Hold on. Pixar's animators take acting lessons? And this was made possible because of the computers? So, it turns out, computers are part of telling the story after all. The leap of computation has allowed Pixar not just to create 3D animation, but also to focus on the story and the way the story is told. The 3D environment makes it possible to show emotions in a way that 2D never could. Shrek's face portrays *feelings*, not just expressions. When he walks across the screen he seems *heavy*, not flat. The use of computer graphics allowed Pixar to create movies that are far more sophisticated than what was possible with hand-drawn animation. Computer technology *enabled* Pixar to do things differently. It enabled them to merge computer animation with traditional filmmaking. Two years later, traditional 2D studios were being deserted everywhere.[16]

This would never have happened without the invention of the microchip, arguably the most significant innovation of the past fifty years. Computing power has since doubled every eighteen months and continues to do so. This exponential leap in computation will generate more intersections for two reasons. First, it will not merely let us do the same things faster (which enables directional innovation), it will also allow us to do *different* things, generating possible intersections between traditionally separate fields. Pixar's ability to affect the way it told its story was a direct result of the additional computing power it had.

The second reason is that the leap in computation has also led to advanced communication. The microchip has paved the way for e-mail, the World Wide Web, mobile phones, satellite phones, television, and cheaper phone calls. It has made our world smaller. That means that individuals, groups, and organizations that used to be separate can now easily come together to find intersections between their backgrounds and expertise. This provides opportunities for both small start-ups and established companies. Consider the following story from Mark Tracy, a marketing manager in Cargill's risk management group.[17]

Cargill is one of the oldest companies in the United States and deals mainly with agricultural products across the world. With revenues over $50 billion, it is the largest U.S. private company ever, larger than Procter & Gamble or AOL Time Warner, with operations all over the world. It may not be the first company that comes to mind when you think about innovation. But its CEO, Warren Staley, says, "It's a great advantage being private, with shareholders who understand agriculture is cyclical, returns are lumpy, and not every risk goes our way." Mark Tracy got to see them take such a risk.

When Tracy joined Cargill as a grain trader, he did not even know what a soybean looked like, much less how much one cost. Yet he was thrust into a position where he had to learn about these things in a hurry. "Suddenly eighty-year-old farmers are asking me what the market is going to do," Tracy said as he recounted long conversations in the middle of fields and grain elevators. This was the way he learned about agriculture and what was on the minds of worried farmers.

A couple of years later, Tracy moved to the company's risk management group, entering an entirely different field. Led by David Dines, a Bankers Trust alum, the small group consisted of ex-bankers who sold complex, customized financial derivatives to huge *Fortune* 500 food customers. These types of companies buy millions of dollars of agricultural products every month and need to protect themselves, or hedge, against potential changes in food prices. The risk management group helped them do just that. The thing is, Tracy realized, farmers face the exact same risk of price changes. After all, they must sell whatever gets bought. "These days, farmers are expected to be expert meteorologists, agronomists, and environmentalists. Oh, and by the way, they have to be expert traders, too," Tracy noted. There seemed to be an excellent opportunity to combine his understanding of the grain business with the risk management group's knowledge of derivatives and help serve the farmers' need.

This intersectional idea could not easily be executed, though. Compared to large corporations, farmers are a diffuse and spread-out group, and at the time many knew nothing about financial derivatives. Reaching every farmer and delivering a customized solution in a language that they could easily grasp was a huge challenge, even though the potential market was enormous. The Internet solved these problems: Although it would have been prohibitively expensive to reach the farmers one by one, the Internet made it feasible for the group to market, communicate, and aggregate risk with farmers at far less cost. Two separate worlds, the new one of complex customized financial derivatives and the old one of grain trading, connected at the Intersection, and thanks to the leap in computation, Cargill could apply those ideas worldwide on a daily basis.

Taking the Next Step

THESE THREE FORCES—the movement of people, the convergence of science, and the leap of computation—are giving rise to more intersections than ever. We live in a world where Colombian artists combine the sounds of the Middle East and the United States; where goat milk, spiders, and fishing lines all have something practical in common; and where we can read a monkey's mind because of the efforts of an interdisciplinary team.

Of course, not all of us want to innovate, and even if we want to, we can choose to stick to one field. But understand this: Because the effects of these three forces are so pervasive, your understanding of a field is likely to become intersected many times during your lifetime. The individuals or teams who find these intersections are likely to be the ones who radically change our world. Yes, we live in an interconnected world, but there is someone making the connections. It could be you.

PART TWO

Creating the Medici Effect

three

Break Down the Barriers
Between Fields

SEA URCHIN LOLLIPOPS AND
DARWIN'S FINCHES

IN EARLY JANUARY of 1995, Jan Sandel, the executive
chef at the Swedish restaurant Aquavit in New York City,
unexpectedly died of a heart attack. The owner, Håkan Swahn, immediately had to find someone to head up the kitchen. He decided to
place newly hired Marcus Samuelsson in charge while he searched for
a permanent replacement. But Swahn was hesitant because Samuelsson was quite young. "Our organization was big and complex, and our
reputation was excellent. It is not the type of operation you just hand
over to a twenty-four-year-old," he explained. In retrospect, it may have
been the best decision he ever made.

At the time, Aquavit had become a well-respected Manhattan
restaurant, with one star from the *New York Times*. But something
strange started happening only weeks after Samuelsson headed up the
kitchen. New dishes based on unique combinations of food from all

over the world began showing up on the menu. The new items, such as oysters with mango curry sorbet, didn't always seem to make sense, but they tickled both the imagination and the palate. They were unlike anything the guests had ever tasted before.

Only three months later Ruth Reichl of the *New York Times* gave the restaurant a rare three-star review because of its innovative and tasty food.[1] Samuelsson was the youngest chef to have ever received such a prestigious rating. "Mr. Samuelsson's cooking is delicate and beautiful," she wrote. Since then he has become known as one of America's leading chefs. He has been featured in magazines such as *Gourmet, Food & Wine, Forbes*, and *Gear* and on networks such as the Discovery Channel and CNN. His cookbook was voted the Best Cookbook in North America, the James Beard Foundation awarded him Best Chef in New York City, and he was recognized by the World Economic Forum in Davos, Switzerland, as one of the Global Leaders of Tomorrow.[2] When Aquavit owner Swahn met Tom Zagat, of the famous restaurant guide *Zagat Survey*, Zagat remarked, "You have become an institution."[3]

What was behind Marcus Samuelsson's spectacular achievements? What were the reasons for his innovative success? Talking to Samuelsson, one might get the impression that pure charm, youthful energy, and hard work are the secret. His voice is filled with vigor and purpose. He is quick to jump up and greet any customer he recognizes, which is almost all of them.[4] His memory of faces and names seems limitless. Within minutes he had me engaged with a number of guests who had just walked through the door. "Meet Renee," he said with a smile. "She is the president of the Swedish-American Chamber of Commerce. You guys should talk." Charisma, energy, and persistence—without a doubt these qualities will help anyone, but they alone cannot explain his rise to chef stardom. Solving that mystery must start with his culinary creations.

There is clearly something special about the food that Samuelsson creates. The menu says the food is Swedish, and you can instantly see that this is true. Ingredients such as herring, lingonberry, and salmon in part define Swedish cuisine. At Aquavit, however, these ingredients are combined with foods you would never see at a typical Swedish restaurant, at least not until Samuelsson began using them. Take a look at the following menu items:

CARAMELIZED LOBSTER
Seaweed Pasta, Sea Urchin Sausage and Cauliflower Sauce

SALMON PLATE
Gravlax and Tandoori Smoked Salmon, Espresso Mustard Sauce and Dill Foam

CHOCOLATE GANACHE
Bell Pepper and Raspberry Sorbet and Lemon Grass Yogurt

Lobster is Swedish; seaweed pasta is not. Raspberry sorbet is Swedish; lemon grass yogurt . . . well, most Swedes at this time had probably not even *heard* of lemon grass, let alone yogurt made out of it. In these recipes we can find at least part of the answer to the mystery of Samuelsson's success. Although it defies intuition, combining tandoori spices and smoked salmon works extraordinarily well, and that daring is what makes Samuelsson unique. Impossible combinations are original and playfully wonderful. How about nettle soup with a sea urchin lollipop? Or a dessert of green apple sorbet with white chocolate mousse and whipped fennel cream? By using Swedish culinary building blocks consisting of seafood, fresh ingredients, game, and certain preservation techniques, Samuelsson combines foods from all over the world, giving Aquavit guests a unique and stellar adventure in tastes and flavors.

Samuelsson has accomplished this by breaking down traditional barriers in cooking. He has an uncanny ability to draw associations from almost any cuisine in the world and see how they connect with his base of Swedish ingredients and cooking techniques. This ability has placed him at the intersection of Swedish food and global tastes. The solution to our mystery now seems rather simple. Samuelsson's creative genius lies in his ability to generate unique food combinations that surprise the palate. He creates food that is daring, distinctive, and, of course, extremely good. Marcus Samuelsson and Aquavit *should* be doing well.

But New York City is made up of thousands of restaurants, many of them with outstanding chefs who have seen and experienced food from all over the world. How was Samuelsson, at such a young age, able to so stun food critics and lay diners alike? How did he escape the limitations of what could be labeled Swedish or European cooking? What enables him to so freely connect disparate concepts, ideas, ingredients, and styles?

The answer is that Samuelsson has low associative barriers. He has an ability to easily connect different concepts across fields. Specifically, he has an ability to find winning combinations of foods from Sweden and the rest of the world. We can all break down our associative barriers like that. In fact, if we wish to find the Intersection, it is a requirement.

What Are Associative Barriers?

TAKE A MOMENT to consider the following situation:[5] Susan is twenty-eight years old, single, outspoken, and very bright. She majored in biology and minored in public policy. As a student, she was deeply concerned with issues of sustainable development, global warming, and overfishing, and is politically active. Which statement is most likely to be true?

A. Susan is an office manager.
B. Susan is an office manager and is active in the environmental movement.

If you answered B, you are in good company; most people would give that answer. But the correct answer is A. If you are confused about this, consider another analogous question. Which statement is more probable?

A. An apple is green.
B. An apple is green and expensive.

This time the answer is apparent; clearly it is more likely that an apple is just green than that it is both green and expensive. The two questions are similar, but expressed in different ways. Yet we tend to make a mistake in the first case but not in the second. Why? The key difference between the two presentations is that in the first case our mind quickly makes a number of associations. Key words, such as sustainable development, global warming, and overfishing, are all associated with the environment. In most instances it would make sense to infer that Susan is active in the environmental movement. Therefore we are more likely to make assumptions about who Susan is as a person, rather than maintain a mind open to possibilities. These connections happen automatically and subconsciously. The effect is subtle, but very powerful.

Psychologists have an explanation for what happens during this process: They say that the mind unravels a chain of associations. By simply hearing a word or seeing an image, the mind unlocks a whole string of associated ideas, each one connecting to another. These chains of associations tend to be clustered around domains related to our own experience. When a chef sees a cod in a fish market she may think of a particular recipe, which in turn makes her think of the menu items for the upcoming evening. But a writer for a sport-fishing magazine may see something very different. He may think instead of his latest fishing trip, instantly recalling the tackle he used and a story he should write about it. The mind works this way because it follows the simplest path—a previous association. Although the chef may know of sport fishing, and even have done it on occasion, it is much more likely for her mind to quickly lead the thought pattern, with little or no effort, to the field she uses most—cooking. Chains of associations are efficient; they allow us to move quickly from analysis to action. Although chains of associations have huge benefits, they also carry costs. They inhibit our ability to think broadly. We do not question assumptions as readily; we jump to conclusions faster and create barriers to alternate ways of thinking about a particular situation.

Researchers have long suspected that these associative barriers are responsible for inhibiting creativity.[6] Experiments have been conducted to examine the difference between high and low associative barriers. One of the first conclusions made by one of the earliest creativity researchers, J. P. Guilford, is that creative minds tend to make unusual associations because they engage in so-called divergent thinking.[7]

Consider the following exercise: What words do you think of when you read the word *foot*?[8] The most common response by far is *shoe*, followed by *hand*, *toe*, and *leg*. Eighty-six percent of the subjects in a test with more than eight hundred people answered with one of these words. On the other hand, only one person each responded with *rat*, *snow*, *physics*, *dog*, or *hat*. Consider another example—what words do you think of when you read the term *command*? The most common responses to that word were *order*, followed by *army*, *obey*, and *officer*. These answers accounted for 71 percent of all responses. Only one person each answered with words such as *polite*, *obedience*, *war*, and *hat*. Guilford's conclusion was that a person with low associative barriers is more likely to think broadly when responding to a word such as *foot* and is therefore able to come up with more unusual ideas. This means that a person with low associative barriers would find his chains of association taking irregular paths outside of a specialized field, rather than predictable ones inside a field. For such a person, *foot* and *command* may even connect; notice that the word *hat* appears in both cases. Individuals with high associative barriers would more than likely produce the common responses, but remain unable to see how the two words are linked unless specifically prompted to find a connection.[9]

This is what I mean when I say that Marcus Samuelsson has low associative barriers. He makes unusual associations outside the field of Swedish cuisine. When Samuelsson thinks of, say, tomatoes, his associations reach further than for most Swedish or European chefs. When I say pesto, he doesn't think basil; he says dill. If I say tandoori, he doesn't instantly think chicken; he says smoked salmon. This can go on all day.

"Lingonberry?" I ask.

"Chutney," he answers.

"Caesar salad?" I suggest.

"Caesar salad soup," he responds.

See what I mean? Samuelsson looks for related concepts in distant places and unexpected areas of cooking and then tries to reconcile these far-flung ideas into recipes. He has, in other words, managed to break down the associative barriers between different fields of cooking. And as a result, his ideas stretch exponentially farther.

How Associative Barriers Help and Hinder Us

IN THE SEARCH for intersections, low barriers provide an advantage. The problem is that there are strong benefits to keeping our natural cognitive barriers in place. Our brain evolved the way it did for a reason. It generally enjoys finding order in things, grouping concepts together, and finding structure in the environment surrounding it. A person with high associative barriers will quickly arrive at conclusions when confronted with a problem since their thinking is more focused. He or she will recall how the problem has been handled in the past, or how others in similar situations solved it.

A person with low associative barriers, on the other hand, may think to connect ideas or concepts that have very little basis in past experience, or that cannot easily be traced logically. Therefore, such ideas are often met with resistance and sentiments such as, "If this is such a good idea, someone else would have thought of it." But that is precisely what someone else would *not* have done, because the connection between the two concepts is not obvious. Two people or two teams—one with high barriers, the other with low barriers—will approach a similar opportunity in completely different ways. Consider the following story about Charles Darwin and John Gould.[10]

When Charles Darwin returned from a five-year trip around the world on the HMS *Beagle*, he had collected a host of birds from the Galapagos Islands. Although Darwin generally was an excellent note taker, he had kept poor records on the birds. The original purpose of his trip, after all, had been to study geology. Once in London, Darwin gave his collection of poorly labeled birds to one of the most prominent zoologists at the time, John Gould. Darwin explained to him that the collection consisted of mixed birds such as finches, wrens, and blackbirds, and they were of little importance to him.

Six days later he heard back from Gould and was surprised to learn that the birds were not such a jumbled mix after all. Gould explained that they "are a most singular group of finches, related to each other in . . . form of body and plumage: there are thirteen species in all. . . ." This confused Darwin. The beaks in these finches were different and used for different things. Some were good for cracking nuts, others for pecking out insects. And then there was the fact that the number of species matched the number of major islands in the Galapagos . . . thirteen. Soon Gould surprised Darwin yet again. Darwin had also collected mockingbirds on the Galapagos Islands, and he had assumed that they were all different varieties of the same species. Gould told him that, no, this was not the case. Instead, each variety represented a distinct species, one from each island. But this was as far as Gould went.

Gould was clearly the expert taxonomist, but it was Darwin who proposed the radical notion: Was it possible for a species of birds to split into two (or more) species if the birds were isolated on separate islands? This notion eventually became the basis for what may be considered the most significant scientific revolution of our time, the theory of evolution.

What is remarkable about this story is not the insight and success that Darwin ultimately garnered, but that John Gould was unable to achieve it. He had the expertise, he was a leader in his field, and he had all the pieces of information available to him. But Gould associated everything he observed according to the rules of taxonomy, and he therefore attempted to fit what he saw in Darwin's bird collection into those rules. His insight was good and helped increase our understanding about the number of finches in the world. Darwin's insight, on the other hand, explained why the field of taxonomy exists in the first place. He had this flash of insight because he was able to break down his associative barriers. The next chapter will show you how to do the same.

four

How to Make the Barriers Fall

HEATHROW TUNNEL AND
RESTAURANTS WITHOUT FOOD

B REAKING DOWN our associative barriers is the first
challenge we face in our search for the Intersection. But
how do we do it? The examples set by Marcus Samuelsson, Charles
Darwin, and others can help us understand. In essence, these people
succeeded at breaking down their associative barriers because they did
one or more of the following things:

➤ Exposed themselves to a range of cultures
➤ Learned differently
➤ Reversed their assumptions
➤ Took on multiple perspectives

Expose Yourself to a Range of Cultures

O NE DAY, while walking in the tunnels from Heathrow Airport to the London Underground, I noticed a prominent advertisement by HSBC, one of the world's largest banks. The ad immediately caught my eye, because it covered the walls all the way from the airport to the Tube. It consisted of several sets of three images. One of the posters showed three identical images of yellow squares. The first square was labeled *USA* and below it was the word *cowardice*, indicating that this was what the color yellow represented in that country. The next yellow square was labeled *Malaysia* and beneath it was the word *royalty*; the last one was labeled *Venezuela*, followed by the words *lucky underwear*.

A bit farther down was another poster showing three identical images of a grasshopper. One image was labeled *USA* and subtitled *pest*, the middle image was labeled *China* and subtitled *pet*, and the last one was labeled *N. Thailand* and subtitled *appetizer*. You get the idea.

HSBC, or the agency representing it, had placed at least ten variations of this ad along the walls. The bank was, rather cleverly, making the point that although it was a global institution, it was also privy to local knowledge and customs. For our purposes, the ad also drives home another point, one that is crucial in understanding how to break down associative barriers: There is always another way to view things. This is particularly true as one compares cultures across the world.

Cultures are defined by rules and traditions. They impose certain ways of thinking and acting. Some cultures are highly social, others are quite reserved; some emphasize teamwork, others focus on individuality. In some cultures spirituality is important, while in others only secular ideas are promoted. One can argue forever whether some norms are valuable all of the time, but one can be quite sure that all norms are valuable some of the time—otherwise they would never have become norms. This is why cultural diversification is so effective in breaking down associative barriers. Through diverse cultural backgrounds and experiences, one can more easily escape imposed viewpoints.

Donald Campbell, one of the leading psychologists in creativity research in the sixties, concluded that "persons who have been uprooted from traditional cultures, or who have been thoroughly exposed to two or more cultures, seem to have the advantage in the range of hypotheses they are apt to consider, and through this means, in the frequency of creative innovation."[1] The point is not that a person who has been exposed to multiple cultures can simply fall back on two or more different ways of viewing an issue. Rather, it is that such a person is not wedded to a particular point of view. Simply by being aware that there are multiple ways of approaching a problem, he or she will more likely view any situation from multiple perspectives.

Cultural diversity does not only imply geographically separated cultures. It can also include ethnic, class, professional, or organizational cultures. The mere fact that an individual is different from most people around him promotes more open and divergent, perhaps even rebellious, thinking in that person. Such a person is more prone to question traditions, rules, and boundaries—and to search for answers where others may not think to.[2] Research also indicates that people who are fluent in multiple languages tend to exhibit greater creativity than others. Languages codify concepts differently, and the ability to draw upon these varied perspectives during a creative process generates a wider range of associations.[3]

For Marcus Samuelsson, the role of cultural diversification was critical to breaking down his associative barriers. For starters, Samuelsson does not look like or have the same history as your typical Swede. He comes from Ethiopia. He was born in the capital Addis Ababa, but was orphaned at three when both of his parents died in a tuberculosis epidemic. His life might have looked very different had it not been for a couple in Gothenburg, Sweden, who decided to adopt Marcus and his sister. Growing up black in Sweden gave Samuelsson the advantage of viewing the world differently from those around him. "I never saw

Gothenburg as my be-all and end-all," says Samuelsson, "unlike most of my friends, who all planned to stay in the area."

He was also fortunate to have the opportunity to visit many other parts of the world as a child. Samuelsson's adoptive father, a geologist, traveled with the children often. Those trips gave Samuelsson early exposure to the breadth of the culinary universe. "When I was a kid, I ate in Poland, Berlin, Russia, and Yugoslavia, and on vacation we'd eat in France, Spain, and other countries. So at an early age I was eating 'weird' food, but it seemed natural."

At sixteen, he entered culinary school in Gothenburg. That in turn led to apprenticeships in Switzerland and Austria. There he learned to speak French and German while also speaking English. Swedish was hardly spoken at all. The most significant of all experiences during his youth was working on a year-long cruise around the world. His description of that trip is a perfect account of how to set yourself up for lowering your associative barriers.

> I got the opportunity to travel around the world on a cruise boat and to eat and cook food at every port. Up to this point I thought good food was "owned" by Europe and France. But during my travels I understood that good food exists everywhere. You have it in Sweden, France, and Switzerland for sure, but even more in Thailand, Japan, India, in Africa, and in South America. That year was probably the most important in my career. We could set out from Öresund [Sweden] one day, three days later arrive at Helsinki [Finland], six days later we were in Amsterdam, ten days later in Bordeaux, twelve days later in Morocco. We went to North America, Brazil, the Amazon, Panama, San Francisco, then the Pacific Rim . . . constantly on our way to someplace new. That was when I realized that if I combine my knowledge from Europe with the tastes that exist in Thai food or Japanese or Latin American food or whatever, then I will have something exciting.[4]

At the end of that pivotal trip, Samuelsson realized that it was time for him to apply his unique perspective and experiences. After working for the three-star George Blanc restaurant in Paris for a year, he ended up at Aquavit in New York. Samuelsson needed an environment where he could focus on cooking without having to explain to restaurant owners or customers that he could indeed cook European food even though he didn't look European. Håkan Swahn, Aquavit's owner, had an open mind about Samuelsson, but admits that most of his American peers would probably have hesitated appointing a black man to head the kitchen of an upscale restaurant that served a distinctly European cuisine. Yet this openness to diversify has become a critical component in Aquavit's operations. The first thing Samuelsson did was to retool the staff makeup, even sacrificing experience for an open attitude. Take one look at Aquavit today and you will see all kinds of people in the kitchen. Aquavit's staff of about one hundred comprises as many as forty nationalities.

By living and working in different cultures and spending significant time learning to appreciate them, one can more easily break down associative barriers or even avoid building them in the first place. Remarkably, Samuelsson's background hits on almost every point of cultural diversification researchers say helps open a person to unusual associations. It has given him an ability to see things that are often missed by others. "Most people confuse the notion of 'Swedish,'" he says. "Sweden today is international and mixed. Sweden today means sushi, rolled by a black guy, served to a Korean couple."

Why not?

Learn Differently

PAUL MAEDER is the founder of the highly regarded venture capital firm Highland Capital. Over the years he has done very well for himself by betting on small firms that ended up becoming extraordinarily successful. Paul Maeder is also a very well educated man. He earned his undergraduate degree at Princeton University, his master's in mechanical engineering at Stanford, and his M.B.A. at Harvard Business School. With all of these degrees, you would think he places an extraordinarily high value on education. But only seconds into our conversation he started listing individuals and groups who have radically innovated because they did *not* have formal training. "Take this guy Stan Lapidus," he told me one day. "He doesn't have an M.D. or a Ph.D., but he has come up with an amazing way of analyzing stool samples for colon cancer pathology. Put it in a blender, mix, and you can spot cancer with hardly any false positives. It's really an amazing invention. Now, why did he think of this? Because he's not a doc."[5]

I am not saying that Maeder thinks education is a bad idea; he would be a walking contradiction if he did. But he clearly sees it as potentially limiting creativity. Why is that? Through school, mentors, and organizational cultures, education tends to focus on what a particular field has seen as valid. If, for instance, you wish to be a great medical doctor, there are rules that must be mastered. A good education will teach you these rules. You learn what past experts and thinkers concluded and use their experiences to build your own expertise. You do this to learn what works. Expertise in a particular field could suffer if too much time were spent questioning basic assumptions. The price for such an approach, however, is that one more easily becomes wedded to a particular way of doing things. As a result, associative barriers are erected, making intersectional ideas less likely.

How do you counteract this effect? One way is to avoid schools and ignore experts. But this would be incredibly impractical advice. Skipping school or dismissing people with valuable expertise makes little sense. Instead, we must employ tactics that allow us to *learn as many things as possible without getting stuck in a particular way of thinking about those things*.

Paul Maeder may have the answer. He has evaluated thousands of business plans and met hundreds of entrepreneurs over the years. The teams that attract him have almost always stepped into the Intersection. "Look at bioengineering, look at materials science. They involve many disciplines, they are inherently interdisciplinary," he says, and starts ticking off dozens of intersectional innovations. "One guy figures out the composition of a new material, another how that material can make better ski bindings. Put them together . . . you've got something." Maeder abhors "single-disciplinary incrementalism" and is always looking for those people who push beyond a field's boundaries.

What, then, does Maeder think are some important aspects of innovative people at the Intersection? Over the years he has spotted two recurring characteristics. "Innovators are often self-taught. They tend to be the types that educate themselves intensely," he says, "and they often have a broad learning experience, having excelled in one field and learned another." Broad education and self-education, then, appear to be two keys to learning differently.

The whole idea behind a broad education, one that covers several fields, is that it can help us break out of the associative boundaries that expertise builds. But is there any evidence that expertise limits creativity? In 1995 the psychologists Robert Sternberg and Peter Frensh set up a study to explore precisely this question under controlled laboratory conditions.[6] In the experiment both experts and novices were asked to play against a computer in a game of bridge. The first round consisted of a standard game. Here the experts clearly played better than the novices. This is, after all, why we call them experts.

The researchers then made some superficial changes in the rules of the game. They changed the ranking (for instance, diamonds rated higher than clubs, rather than the other way around) and the names of the suits. The players suffered momentarily from the changes, but quickly recovered. All they had to do was to learn the new rankings or names. Again the experts played better than the novices.

The interesting part of their experiment occurred during the deeper structural changes. In bridge, after the cards have been doled out, there is a bidding phase, followed by a playing phase. The playing phase occurs in a number of successive rounds. Normally the player who puts out the high card in a given round wins and leads off play in the next round. But the researchers reversed the rules for the playing phase so that the player who puts out the *low* card wins. The effect of this change was hardly noticed in the novices' performance. They did not need to disrupt complex strategies of play since they had never developed them in the first place. But that was not true for the experts. They could no longer use their strategies and had difficulty coming up with new ones. In other words, expertise, for all its strengths, can make it more difficult to break out of established patterns of thought.

John Donaghue, the director of the Brain Science Program at Brown University, agrees. He believes that the school's ability to integrate undergraduates with the work of graduate researchers and professors could be a huge advantage. Obviously the students themselves benefit from such a system. But so does the research team, Donoghue explains. "They have different ideas, ideas that we have become too blind to see. Many of these ideas turn out to be very good." This is not to say that younger people are more creative. However, younger people are often less constrained by their education within a field since they have not yet had too much of it. It would follow, then, that learning a new field, whether one is young or old, can help break down associative barriers. Thomas Kuhn points out in his seminal book *The Structure of Scientific Revolutions* that "almost always the men who achieve . . . fundamental inventions of a new paradigm have been either very young or very new to the field whose paradigm they change."[7]

Paul Maeder's second characteristic for success at the Intersection was self-education. By learning fields and disciplines on our own we have a greater chance of approaching them from a different perspective. In fact, formal education often looks like an inverted U when correlated with one's success as a creator. That is, formal education first increases the probability of attaining creative success, but after an optimum point it actually lowers the odds. This point occurs a bit earlier for artistic careers and a bit later for scientific paths.[8]

There are numerous examples of this. Thomas Edison, probably the greatest inventor ever, did not achieve any higher levels of education.

He was, however, a voracious reader of anything and everything that interested him. By the time Edison was twenty years old he had read most major books on chemistry and electricity and conducted hundreds of experiments based on what he had read. He would say that books could "show the theory of things," but that it was "doing the thing itself that counts."[9]

Here's another example: Steve Jobs, the founder of Apple and Pixar, did not complete college. This is not to say that he did not educate himself—just not at school. All of this suggests that it makes sense to spend significant amounts of time reading and drawing, learning and experimenting, without guidance from instructors, peers, and experts. It is ironic, then, that many people who wish to innovate find that they do not have time for such side ventures. But if innovation is the goal, such experimentation is precisely what one must aim for. Charles Darwin was a below-average student because much of his time was filled pursuing botanical interests in the English countryside or conversing directly with established scientists. Darwin's father lambasted him for not showing an inclination toward any one thing in life. First Darwin enrolled to become a physician, then a minister. Neither bore out. Finally he decided to board a boat, the *Beagle*, for a five-year trip around the world to study geology, essentially on his own. He ultimately became the most significant biologist, and possibly scientist, of all time. Darwin concluded, "I consider that all that I have learned of any value to be self-taught."[10]

Reverse Assumptions

THE TWO STRATEGIES discussed so far involve long-term approaches to breaking down barriers between fields. But this does not help us much if we need some fresh insights *right now*. Is it possible to forcibly break down the associative barriers when confronted with a particular challenge? Can we, in other words, actively search for the Intersection? There is significant evidence suggesting that we can.[11] We often hear advice that we should "unlearn" what we have learned, or ignore the experts around us, in order to free our minds. Such recommendations can be frustrating. Although they make sense on one level, they provide precious little guidance for execution. How do we rationally ignore experts or unlearn what has worked for us in the past?

Forcing a breakdown of associative barriers means directing the mind to take unusual paths while thinking about a situation, issue, or problem. One of the most effective ways of accomplishing that is to perform an assumption reversal. By reversing assumptions the mind is encouraged to view a situation from a completely different perspective, clearing the path to the Intersection. Perhaps the single most significant discovery for making commerce possible on the Internet came from an assumption reversal.[12]

During the more than 2,500 years that codes and encryption have been used, one basic "law" has always ruled: In order for one party to encrypt a message and another party to decrypt it, both parties must have the same code key. An analogy to this law is that if I place a secret message in a box and place a padlock on it, you can only open the lock with a duplicate key, which I must have given to you beforehand.

The effect of this "law" on Internet commerce would have been devastating. Imagine if you first had to agree with the online bookseller Amazon.com on a code key before entering a credit card number on its site. This key would have to be delivered in a way that no one else could access; e-mail, for instance, would be far too risky. You could meet locally with a representative of the company, but this would obviously defeat the benefits of purchasing the book online in the first place. The fact that both parties need the same key could have stalled the entire commercial development of the Internet. Fortunately, this did not happen.

In the early seventies, when the Internet was in its infancy, two brilliant code breakers at Stanford University, William Diffie and Martin Hellman, reversed the most basic of all assumptions in cryptology.

What if both parties did not need the same key? Such a proposition seems to defy logic; how would that even be possible? But it was. By reversing this assumption, Diffie and Hellman found the intersection between the field of cryptology and a particular, curious brand of mathematics involving so-called one-way functions. The best way to mentally understand how these functions work in cryptology is to return to the box example. Imagine that a person, let's call her Alice, has a padlock. She can give a copy of this padlock to anyone who asks for it. So when you wish to send a message to Alice, you ask for her padlock. After you get it, you stick your message in a box, lock it with her padlock, and send it to her. Once you lock the box, however, even you can't retrieve its message. Only one person can do so, Alice, because she has the only key around. Later, three other researchers at MIT, Ronald Rivest, Adi Shamir, and Leonard Adleman, made this type of cipher commercially viable, and it has become known as the RSA cipher. Without it you would not be able to purchase anything securely over the Internet.

Assumption reversals are a remarkably effective way to challenge the way you think about almost anything. The example outlined here comes from the outstanding book *Cracking Creativity* by Michael Michalko.[13] The purpose is not necessarily to come up with a specific idea, but to shake your mind free from preconceived notions. This is how it works:

1. First, think of a situation, product, or concept related to a challenge you are facing, and think about the assumptions associated with that situation.
2. Next, write down those assumptions; then reverse them.
3. Finally, think about how to make those reversals meaningful.

For instance, suppose you wish to open a new restaurant but are having difficulty coming up with a novel concept. First list some of the more common assumptions involved in running a restaurant, and then reverse them. Your list could look something like figure 4-1.

FIGURE 4-1

Reverse Your Assumptions

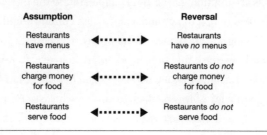

Assumption		Reversal
Restaurants have menus	◄········►	Restaurants have *no* menus
Restaurants charge money for food	◄········►	Restaurants *do not* charge money for food
Restaurants serve food	◄········►	Restaurants *do not* serve food

Now try to think of ways you could conceivably build a sustainable business out of each reversal. Here are some examples:

➤ *A restaurant with no menus:* The chef informs each customer what he bought that day at the meat, vegetable, and fish markets. The diner selects the desired food items and the chef creates a dish from them, specifically for each customer.

➤ *A restaurant that does not charge for food:* This restaurant is a café where people get together to talk and work with each other. The café charges for time spent instead of food consumed. Selected low-cost food items and beverages are given for free.

➤ *A restaurant that does not serve food:* The restaurant has a unique and beautiful décor in an exotic environment. People bring their own food and beverages in picnic baskets and pay a service charge for the location.

If a solution seems particularly attractive, you can keep elaborating upon it, thinking openly about how to make it happen. The point is not to immediately find the solution you are looking for (although that could happen), but to put aside, at least temporarily, the most obvious assumptions and allow your mind to escape its usual chains of association.

There are other ways to perform assumption reversals. You can, for instance, take a goal, reverse it, and then try to figure out how to achieve the reversed goal. Such a process would force your mind to approach a common topic in an unusual way. Consider this question: How could you make the banking experience as pleasant as possible?

Chances are you may have pondered this question before, even if you are not a banker. You can almost feel your mind starting to walk down a familiar path. Answers that come to mind might include friendly customer representatives, appealing décor, and convenient locations for ATMs. But what would happen if you reversed the goal? How would you make the customer experience as horrible as possible? How could you drive customers *away*? It is unlikely that many people have seriously considered this question, but the answers might yield some interesting and unique insights.

Try on Different Perspectives

I MAGINE WATCHING a flower grow through time-lapse photography. It's the type of jerky film motion you may have seen on nature programs that show a flower sprout from the ground, bloom, wither, and die within seconds. Such a film helps us understand the full cycle of the flower and gives us a perspective on its life.

Now, change your perspective. Instead of observing the flower, *become* the flower.[14] Imagine being a camera inside the flower with the ability to record the surrounding environment. The camera would record the weather, the rain, the ground; it would film the nutrients as they move through the soil and up the roots; it would record the water, the gardener tending the flower, and the bumblebees pollinating it while seeking nectar. This perspective will give you an entirely different set of insights about the nature of the flower—perhaps unusual ones, because it is a different way of looking at something common. Which perspective will give you fresh insights? Which will elicit more ideas about the flower from a scientific standpoint, and from an artistic standpoint?

We can choose how to view any situation. If we always view it from the same perspective, we will tend to notice the same things. Look at the figure. If asked, most people would probably describe it as a square composed of alternating rows of circles and triangles.

It isn't as obvious that the figure consists of alternating columns of circles and triangles. If we just ask different questions about a problem, we can see it in a new light, and possibly engineer a breakdown of associative boundaries. Leonardo da Vinci, the defining Renaissance man and perhaps the greatest intersectionalist of all times, believed that in order to fully understand something one needed to view it from at least three different perspectives.[15]

One of the more radical innovations in environmental management and control emerged by viewing an old problem from a different perspective. During the seventies and eighties, environmental groups and industry viewed air pollution and the resulting acid rain as an ecological or political problem, respectively. This led to legislative battles and loophole-ridden policies. One major innovation for effectively dealing with air pollution came about in 1990 when politicians and environmentalists began to view the problem from a market-based perspective. By organizing a marketplace where companies could trade pollution rights, the overall level of emissions dropped by greater levels than ever before. Such an approach has since been emulated by other nations for this and other environmental problems around the world, such as global warming.[16]

You can view a situation from any number of perspectives. So why always choose the one that comes the easiest? By forcing yourself to view a project differently, you can break down associative barriers between fields and uncover unexpected connections. This sounds a lot easier than it actually is, of course. To make it work you must choose perspectives that are radically different from the ones you usually work with. Once again, as with assumption reversals, the point is not to come up with a specific idea per se, but rather to free up the mind and escape the routine chains of association. Here are a couple of suggestions:

> *Apply the idea to someone or something else*: Imagine that you are designing a beach house. What would it look like? Now assume that you are designing that house for Pablo Picasso— how would that change the design? Forget that you have no idea of what he actually wanted, but work from your perception of who Picasso was as a person. Then suppose you were designing the house for opera singer Luciano Pavarotti. What would happen to the size of the rooms, the curvature of the valves? The ideas you would get from these types of explorations could evolve into something interesting and unique when combined with your standard way of thinking about such a project.

> *Create constraints:* When a yoga teacher broke her arm, she was not sure if she could continue teaching while it healed. She soon found, though, that without the use of her arm, she naturally resorted to new and inventive methods for both understanding her own body and teaching yoga. By creating constraints, by accident or on purpose, we may be pushed to explore alternative ways to solve a given problem. Say that you are trying to innovate your in-store customer service operation. What happens if you assume that the customer service personnel can't speak? Or can't use their hands? By creating constraints, you may break down the barriers and think of ideas that would never have occurred to you otherwise.

What Happens Next?

T HE STORY OF AQUAVIT is a story of success. In fact, Marcus Samuelsson is about to open another restaurant.

"In which city?" I ask, assuming he will expand the Aquavit concept elsewhere.

"Here, in New York," he says. "It will be a Japanese restaurant."

This catches me off-guard. Japanese? But then I understand. After all, who is best suited to innovate the Japanese cuisine? Would it be an expert at cooking Japanese food? Or would it be someone like Marcus Samuelsson?

The stage for Samuelsson had been set long before he arrived in New York. His background, education, and propensity to reverse common assumptions about Swedish cuisine and to view it from different perspectives enabled him to connect culinary concepts from around the world. He found the Intersection because he managed to break down his associative barriers.

That, however, is not enough to innovate. Intersectional ideas consist of combinations of concepts from different fields. How do these combinations occur? And what is the secret behind an idea that really makes it big? We will look at these and other questions in the following chapter, in which a young mathematician manages to take the gaming world by storm.

five

Randomly Combine Concepts

CARD GAMES AND SKY RISES

I N THE SPRING OF 1991, a young Ph.D. math student named Richard Garfield met with Peter Adkison, the president of a small game company called Wizards of the Coast. Garfield had designed a board game called RoboRally and he was pitching the idea to Adkison. But Adkison did not bite. "Come back with something less complicated," he told the mathematician. He suggested that Garfield design a game that was quick to play, portable, and inexpensive to produce.[1]

What Garfield came up with revolutionized the world of games. He created Magic: The Gathering, a card game unlike any other. During the second half of 1993, following the release of Magic, Wizards of the Coast made about $200,000, which isn't bad for a seven-person start-up. The following year, however, that same small company made $40 million, and in 1995, Wizards of the Coast sold over 500 million cards. Magic had launched a gaming epidemic. Ten years later there were more than 6 million Magic players in more than fifty countries and over 100,000 professionally sanctioned tournaments around the world each year. In fact, Magic created an entire genre of games. When Wizards of the Coast launched the Pokemon card game in the United States, its addictiveness among kids all over the world prompted religious groups to denounce it. Wizards of the Coast's success soared. Magic, and the industry it spawned, had become part of our culture.[2]

How did Richard Garfield create such an incredible game? And how did he get from RoboRally, an idea that led absolutely nowhere, to Magic, one that made him a legend almost overnight? In order to unravel these mysteries, we have to understand what occurs after the breakdown of associative barriers. We must understand what actually happens at the Intersection.

Finding Magic

RICHARD GARFIELD is a measured talker.[3] He takes his time to think about a question before answering it. "Still here, still here. I'm just formulating my answer," he says unassumingly during a phone call. His comments are precise, yet also tentative as if he wishes to give a clear answer but still leave some room to revisit it later on. Maybe it was his Ph.D. background in combinatorial mathematics that paved the way for such an exacting nature, or maybe it was his background in game design that kept him open to possibilities. Whatever the reason for his makeup, it is clear that this is someone who loves every aspect of games and gaming.

Magic was Garfield's hobby for a long time. He would keep it on his shelf only to take it out every couple of months to "tinker with it for a little bit, play with my friends perhaps, and maybe test out new rules." Then it went back on the shelf, sitting there until the next session. All in all, he had tinkered with Magic for eight years before it actually went to market, although that only represented a couple of months of real work. But Garfield does not directly credit these eight years for coming up with the idea of Magic. He credits it to one day spent in the country. "Everything about my game making is evolutionary. The one exception to that is Magic. The idea that made Magic into something special came one weekend while I was visiting my folks in Oregon—we had gone to Multnomah Falls. I can remember exactly where it happened and exactly when it happened. I had this Eureka. And the idea . . . the idea seemed to come out of nowhere."

To understand what was so revolutionary about Garfield's idea we must first understand a little about how the game works. In Magic two players face off against each other with their own sets of cards. These cards are divided into categories such as creatures, lands, and spells. The point of the game is to use your cards in various strategic combinations to destroy your opponent by bringing his or her life-force points down from 20 to 0. So far this seems like nothing spectacular. It may remind you of a slightly more elaborate cards version of chess; in both games you can develop multitiered strategies with pieces that have different functions.

But Garfield's idea at Multnomah Falls gave Magic a crucial design difference; one that made it distinct from virtually all other games that had preceded it. "The great breakthrough with Magic was when I realized that not all the cards had to be the same for all people," Garfield recalls. Before a game starts, each player assembles a deck of sixty cards by balancing monster cards, landscape cards, and spell cards. These sixty cards come from the player's private collection. One player's collection can look very different from another player's because there are hundreds, even thousands, of cards in total circulation.

This is how it works: When a player buys a deck of cards he gets sixty, but those sixty represent only a fraction of the available cards in the entire card set. If the player buys another deck, he will probably get some cards he already owns along with a bunch of new ones. This means that when one player uses, say, a Juggernaut monster card, the other player may never have seen it before. Even so, the other player will quickly understand how this new card affects her own strategy and can therefore easily integrate it as the game keeps going. Because players bring their own decks, they can actually play an entire game with cards that none of their opponents has seen.

Think about that for a minute. Imagine walking into a game of poker where a player suddenly presents a straight flush in a totally new suit of cards. "These are ovals," he says. You would probably get confused (or maybe very angry). Games just don't work that way. For essentially all of history, all the pieces of a game have to be present for fair play. If you play chess, for instance, you expect all the pieces to be on the board in their correct positions before the opening move. Not in Magic.

Once the game is over, the second player may take a closer look at the new Juggernaut card, decide that she likes it, and offer to trade it for one of her own duplicates. It turns out that some cards are common; others, such as Juggernaut, are rare. Rare cards may be difficult to get no matter how many decks you buy, and the only way to acquire them is by trading with other players. This can involve joining a community of players locally or on the Internet, or meeting fellow players at conferences. On top of that, Wizards of the Coast releases new card sets every year, making the card search (and card buying) a continual and fresh challenge.

What is the result? Players buy entire decks simply to get one particular card. Even more interesting, they find a million and one ways to locate other players with whom to trade cards. Soon players begin trading cards for reasons other than to improve game play—perhaps because they predict an increase in the value of rare cards or want to get a complete deck.

Wait a minute. Isn't that what collectibles are all about? Think baseball cards. Think stamps and coins. Remember the Garbage Pail Kids cards? These items can be bought, collected, and traded, leading to an amazing self-reinforcing and rapidly expanding network of collectors.

That, then, is the secret of Magic: The Gathering. It sits at the intersection of collectible items and ordinary games and is called, not surprisingly, a collectible card game (or trading card game). The inter-sectional idea that hit Garfield that day in Multnomah Falls in Oregon was a concept from a field other than games—collectibles—but he connected the two worlds. That connection was both unique and wildly successful. "When the game hit the market it was just incredible how fast it was selling . . . it was spreading like a virus," he says. "When I talked about the game and its rules at conferences, people's attention was rapt, they were intensely immersed. I don't know what was so com-pelling, but I had never seen people so focused on anything before or after. . . . The initial 10 million cards we published were gone in [about] four months."

Garfield offers two reasons for Magic's success: a prolonged and exciting learning phase and an expanding community of players. Ex-amined closely, you will see that he is talking about the intersection of games and collectibles. "The players in any game go through several different stages," Garfield explains. First they learn the rules. After that there is an exciting part of the game where the players learn major strategic ideas. In chess, for example, this may be how to protect pieces. If two people are learning at the same time, the person who dis-covers the next big thing wins; then the second person copies and im-proves upon the new strategy, and this continues back and forth. Slowly, Garfield says, the game enters the third phase, where strategy is much harder to innovate and the rewards are much smaller. Most players find this phase burdensome, and they either fall out of the game or settle into a more comfortable method of play. During this phase of chess, players may keep at it but do not really improve and are essentially playing the same game over and over. "Magic is a bit differ-ent in that this big improvement stage is with you for a long period, since the cards keep changing."

"In addition," Garfield continues, "Magic has really created a com-munity, much more so than a regular card game or board game. When you play this game with your friends, you see that they have different cards than you do, so you start discussing strengths and weaknesses of cards and decks. Some cards might be traded . . . and you become a viable part of this community and get sucked in." In Magic's case this is a very interconnected group. Players will seek out friends of friends, people they may have never met, just to get a specific card. People in this network, Garfield points out, interact in a much more dynamic way than in a game like Monopoly. If you play Monopoly with friends and they like it, they might buy a copy and play with their friends—but that's pretty much it.

It all seems so simple and obvious when Garfield talks about Magic and what happened that day at Multnomah Falls. But if it was that obvious, others would have thought of it. What specifically was behind his moment of insight? How, exactly, do we generate intersectional ideas?

The Creation of an Intersectional Idea

A N EARLY ATTEMPT by the psychologist N. R. Maier to understand the nature of insight has become a well-known creativity experiment.[4] The subject of the experiment is led into a room. There he or she sees two long strings hanging from a high ceiling. Close by is a desk with a variety of tools, including a pair of pliers. The subject is told that the object of the experiment is to tie the two strings together and that he or she can use any of the tools available to solve the problem. Usually the subject tries to first tie the strings by simply pulling them together, but this, as you may have guessed, is not possible. If the subject grabs one string and walks over to the other he or she will find that it is out of reach. The strings are too far apart.

In order to solve this puzzle the subject must use the pliers in an unusual way—as a pendulum. Once the pliers are tied to the end of one string, the person can set it in motion, causing the string to swing back and forth. The person can now pull the second string toward the first one and, when the pliers swing back in their pendulum motion, easily grab them and tie the strings together.

Although the solution may seem obvious in this context, most people find it difficult to solve in real time. The revealing part of Maier's experiment came when he tried to understand what would make the solution more apparent. One factor was the type of tools offered. Using pliers as a weight required the person to think of them in a completely different context, to use them in an unusual manner. But if one of the available tools was a plumb bob, which is used as a weight for pendulums, subjects found it much easier to solve the problem. Maier also found that the subjects of this study responded to hints. In some cases the experimenter would "accidentally" brush against one of the strings, setting it in motion. In those instances the people in the study were much more likely to quickly solve the problem. Interestingly, the subjects were often unable to identify the hint as the triggering factor. When asked what made them think of the solution, they had no idea.

One can draw at least two important lessons from this experiment. The first is that creativity comes from combining concepts in an unusual fashion. Pliers and a string, although separate at the outset of the experiment, become one—a pendulum. The second lesson is that it is difficult to trace the origin of an insight. The triggering factor appears random, lucky, or, as Richard Garfield said, "to come out of nowhere." Creativity, in other words, is a combination of concepts and it is random. Let's examine these two lessons in greater detail, because they are critical to our understanding of how to create an intersectional idea.

Lesson 1: It's a Combination of Different Concepts

Arthur Koestler was the first sociologist to offer a broadly accepted theory for how creative ideas emerge when concepts clash together. Koestler suggested in the early sixties, in his influential book *The Act of Creation*, that the process of creativity is similar to the process that makes us laugh.[5] Have you ever wondered, for instance, why you burst out laughing when you hear a good joke? Or what, exactly, makes a joke good?

If you think about it, a joke is usually a story that starts off along a particular axis, one you quickly buy into. But then, suddenly and unexpectedly, that story is intersected by another concept. The break, or collision of concepts, prompts a reaction—in this case laughter. Consider the following story:

> *Three men have died and are waiting to enter heaven. Saint Peter, the gatekeeper, tells them that heaven is full at the moment and that he will only admit people who have had an unusual death—and therefore asks each of them to explain how they died. The first guy starts:* I live on the fourteenth floor in a sky rise and I suspected that my wife was having an affair. So I came home early one day and searched everywhere for her lover. I finally found him hanging off the balcony, trying to hide, so I rushed out and started hitting him in uncontrollable fury. The lover finally lost his grip and fell, but he was miraculously saved by some bushes on the ground. I tried to find something heavy to throw down on him and ended up heaving the refrigerator on his head. But all the excitement gave me a heart attack, and I died.
>
> *He was immediately allowed in.*
>
> *The second guy followed:* Well, I live on the fifteenth floor in a sky rise and I was cleaning my balcony when I suddenly slipped and fell. Amazingly, I was able to grab hold of the railing below and could see the man inside that apartment come running to my rescue. But instead of saving me he started kicking and beating me. Finally I could not hold on any longer and fell. Incredibly, I survived because of some bushes next to our building. But then this random refrigerator fell from the sky, hit my head, and I died.
>
> *He, too, was let in.*
>
> *Finally, the third guy says:* So, I was hiding naked in this refrigerator . . .

This story has a direction. You may even have started to smile as the implications of the mix-up between the man cleaning his balcony and the suspected lover became clear (directional idea). The story was then intersected by an unexpected concept—that the refrigerator was not filled with food, but with a man. The joke is a vivid example of what happens when people in one field unearth a new insight by combining their knowledge with unrelated ideas from a separate discipline. Often such combinations are followed by an immediate reaction such as laughter, or as Koestler called it, the "haha" reaction. In contrast, artistic originality evokes the "ah!" reaction, and scientific discovery the "aha!" reaction.[6]

These types of moments happen to us all the time; we just have to recognize them. Robert Johnson, for instance, got the idea for Black Entertainment Television, BET, while sitting in a cab listening to someone pitch an idea for a cable TV station targeting the elderly. Realizing that this was "what we're already doing in the black community with print," he suddenly made the connection between cable TV and African American consumers, a connection no one believed was possible in 1979. Twenty years later he sold BET to Viacom for $3 billion.[7]

This also explains why specifically intersectional ideas tend to be so remarkable. One of the earliest creativity researchers, Sarnoff Mednick, wrote, "The more mutually remote the elements of a new combination, the more creative the process or the solution."[8] In other words, if the concepts combined are very different, the new idea will be correspondingly more creative. That's why combining *conch shell* with *beach* will not intrigue anyone, but combining *conch shell* with *armored tank* will. That's why linking *game show* with *money* elicits yawns, while combining it with *reality show* launches an entire genre of TV programming.

Intersectional ideas are groundbreaking, then, because the concepts involved are so different and the combinations so unusual that no one would have thought them possible. Although such combinations do not always lead to anything useful, sometimes they do—and in those cases they can work just like magic.

Lesson 2: It's Random

It is always interesting to hear people talk about when and where they realized a particular insight. Since the trigger is unexpected, it usually makes for a good story. Consider, for instance, Garfield's eureka moment. He spent eight years tinkering with Magic. But the big breakthrough, the one that would bring the game from a personal hobby to a global revolution, came in an instant. And it came while he was at a waterfall. Why then and why there? Although he is an expert at games, there was no specific reason for why he would have thought of the key to Magic at that moment. It seems possible that he could have gone another couple of years without the insight. Did he just get lucky?

Luck does indeed seem to be critically important for innovation. When artists, entrepreneurs, or scientists talk about reasons for their creative successes, you hear them say, again and again, that it was because of luck. (Hard work being the second reason. The relationship between these two factors will become clear in chapter 7.) Malcolm Gladwell, whom I talked about briefly in chapter 1, is not just a book author, but also a writer for *The New Yorker*. In his book and articles he shows an uncanny ability to connect concepts from different disciplines in engaging stories. He connects topics such as suicides in Micronesia with decreasing crime levels in Manhattan, and attacks on police with reading people's minds. I asked him how he comes up with his ideas. "It is very serendipitous," he told me. "Sometimes I don't know how it happens. It is very random. Sometimes someone will say something to me and it is interesting and I will remember it. It is important to be completely open at all times, to be surprised by some piece of information. Half of the time I can't even remember how I came up with an idea."[9] He is not alone in feeling this way. Most people I've talked to have a great deal of difficulty explaining how they generated their insights and why they didn't happen sooner.

Research has shown that two main types of random combinations are involved in generating creative ideas.[10] The first form, which I call "flash-in-the-sky serendipity," usually happens while you are trying to solve a problem. Usually there is a specific goal in mind; you are just not sure what the solution or finished product looks like. This situation is very common in the work place. The goal might be an innovative marketing campaign, a new grant, or a special-effects technology. In these situations the solution tends to emerge after, first, a long period of intensive thought, and then a period where one does not think about it much. During that period the problem is still held persistently in the mind—for hours, days, weeks, months, or even years—while it is temporarily associated with other concepts and impressions acquired accidentally in the course of the day. Sooner or later, one of them "clicks" with the problem at hand and a new idea or solution emerges.

This period, while the mind is "silent" and bombarded with impressions throughout the day, is called the incubation period, and it is a well-documented phase of creativity. Garfield's insight for Magic is a great example of this type of serendipity. These "flash in the sky" moments are not reserved only for highly creative types. We have all experienced times when an idea simply emerges out of nowhere, triggered by the connection of a seemingly unrelated event.

The second form of random combinations, which I call "prepared-mind discoveries," happens when someone with a "prepared mind" encounters a phenomenon he or she had not set out to find. I say "prepared mind" because this particular observation could easily be missed unless one is prepared to understand its significance. A person can be working very hard on something in one area, but then by chance make a discovery regarding something fairly unrelated. Many examples of this type of random discovery are documented in the area of science and technology. The most famous one is perhaps Louis Pasteur's discovery of vaccination in 1875. Pasteur had forgotten a culture of chicken cholera bacteria in his laboratory over the summer. When he came back and injected the old bacteria into the chickens, they didn't die, as expected, but became only slightly ill, and then recovered. At first Pasteur thought there was something wrong with the bacteria, so he got a new culture. When he injected the new culture into the chickens, they still survived. Pasteur suddenly realized that the chickens had been immunized, or vaccinated, during their first injection—a completely unintended discovery! Had he not been prepared to understand the significance of the chicken surviving, however, the insight would have escaped him.[11]

Prepared-mind discoveries are well documented in the sciences, it seems, because science has a tradition of stating a purpose for experiments, a hypothesis. It therefore becomes immediately obvious if a conclusion lies entirely outside the scope of what one is examining. But the same serendipitous process is evident in entrepreneurial ventures and in artistic endeavors. For instance, many start-ups set out to sell a particular product to a particular type of customer, but by the time they make it through their first couple of years, they have often switched products or customer targets based on unexpected, or random, observations of what worked or what didn't. Once again, these types of random events happen to everyone.

Most people are a bit bothered by the notion that creativity is so dependent on chance. We imagine that logic or skill or something else, anything else, should have much more to do with it. We think it should be possible to figure out what is creative, and not be hostage to "flash in the sky" moments or "prepared-mind discoveries." But we are. Consider Magic. Could Richard Garfield have logically figured out the unique combination behind that game? Not likely. That is why he first presented the dud RoboRally to Peter Adkison before showing him Magic. There was no way for Garfield to know that he would have an incredible insight for another game only a couple of months later. Nor could he have known where he would have this insight. Magic, in other words, was the work of luck.

But this is not the whole story. If it were, there would be no practical point in reading this or any other book on innovation. We know that some individuals, teams, and organizations are a lot more innovative than others. If creativity were just a question of randomness, this would seem improbable. Is it possible to increase the chances of finding extraordinary intersectional ideas? It is not only possible; it is essential if you wish to generate groundbreaking innovations. The next chapter will show you how.

How to Find the Combinations

METEORITE CRASHES AND CODE BREAKERS

O NCE YOU HAVE BROKEN DOWN your associative barriers, you will be more open to the random combinations of concepts between fields. Although you may never be able to fully control those combinations, you *can* increase the chances for them to occur. You make that happen by stepping into the Intersection. This chapter will show you how various individuals and teams have done it:

➤ By diversifying occupations
➤ By interacting with diverse groups of people
➤ By going Intersection hunting

Diversify Occupations

O NE OF THE PREVAILING MYSTERIES of the past century has been just what caused the extinction of the dinosaurs. These creatures roamed the continents for millions of years until, quite suddenly, they disappeared around 65 million years ago. The rapid extinction mystified paleontologists for decades. Their speculations were many and included seriously proposed theories that dinosaurs had developed hay fever, that they were outcompeted by emerging mammals, or that they simply became too big.[1] It took Nobel laureate and astronomer-physicist Luis Alvarez to propose that a ten-kilometer-wide asteroid had struck Earth during the end of the Cretaceous period.[2] The asteroid would have kicked up a wide dust belt that would have blanketed the Earth's atmosphere, leading to cooler temperature and ultimately to the demise of an entire branch of the evolutionary tree. This is now the leading theory to explain the great extinction eons ago.

Paleontologists were aware that asteroids and meteorites had struck Earth throughout its history. So why didn't someone from their field propose the asteroid theory? Well, simply put, they didn't think of it. By coming from a different discipline, Luis Alvarez was able to connect astronomy with paleontology, and therefore he had a greater chance of finding an idea that had escaped the other experts.

The act of moving between, or switching, fields through different jobs, projects, or hobbies can be an effective way to generate unplanned, unique insights.[3] I call this process *occupation diversification*, and it is a common way of finding intersections. This is what Luis Alvarez did when he—a man with a background in astronomy and nuclear physics—took an interest in paleontology. Of course, in order for it to work we must be able to associate freely between the different backgrounds, as discussed in chapter 4. If we can manage that, though, we can often transplant old methodologies or frameworks into the new environment and generate unusual idea combinations. Consider, for instance, what happened when an engineer became curious about the long loops we have in our kidneys. For many years physiologists had assumed that the loops had no special function and were a relic of the way the kidney had evolved. But they reminded the engineer of a countercurrent multiplier, an engineering device for increasing the concentration of liquids. And he was right, that is exactly what they are used for in our bodies.[4]

It makes sense, then, to spend time on a variety of projects in different fields if you wish to generate intersectional ideas. Unfortunately, most organizations do not work like that, making occupational diversification difficult to pull off. Usually a company is set up to identify the optimal job for each employee. Once that position or area has been identified, the company then supports further specialization. If you are, say, an expert at grain trading, the company will be hard-pressed to move you to managing health care delivery. You are more useful to the company in grain trading. To move a person from an area he or she excels in to an area the person hardly knows seems to defy common sense. If your goal is to keep execution at a premium and to innovate in small, directional steps, specialization is the right path. However, if you wish to develop fresh, groundbreaking ideas, highly varied experiences are critical.

One firm that understands this principle is Bain & Company, and the firm's chairman, Orit Gadiesh, is the driving force who makes it happen. Bain—one of the world's leading strategy consulting firms—is a company that helps organizations develop innovative growth strategies. If a client wishes to enter the German market with its product line, for instance, Bain can help the firm develop a specialized and successful approach.

Gadiesh is known as the person whose leadership brought Bain out of financial difficulties in the early 1990s and set the firm upon a successful growth path. Her reputation and mystique are well known in the consulting world. So is her history. She spent two years in the Israeli intelligence unit, "learning not to be intimidated by important people." After completing a degree in psychology, Orit Gadiesh left

Israel to enroll at Harvard Business School, while barely knowing a word of English. She graduated two years later in the top 5 percent of her class.

Gadiesh is a rebel, a radical, in many ways. She is blunt with her clients and not afraid to go against the crowd. When everyone was riding the tsunami-sized tidal wave of the Internet boom, she refused to surf along. "It's a tool! Not a paradigm shift," she would say, earning her the distinction of "dinosaur" from pundits at the time. This was, of course, before the collapse of the Internet bubble and the demise of multibillion-dollar companies whose value evaporated and whose names no one remembers today.

When I met Gadiesh at Bain's headquarters in Boston, I noticed right away that she didn't look like a stereotypical consultant.[5] No navy blue dress, no three pieces of jewelry. She looked the way I thought a corporate rebel would look—like herself. She must have worn at least fifteen bracelets, and sky-high heels punctuated her walk. Her smile was disarming and her eyes focused. Less than a minute later we were talking about intersections.

"Some people say that the modern-day Renaissance man is an investment banker who likes to go horseback riding on the weekend he has off, or something like that," she says with a laugh. "That's not a Renaissance man, that's a man with a hobby. A Renaissance man is someone that can see trends and patterns and integrate what he knows. To me the modern Renaissance man is curious, interested in different things. You have to be willing to 'waste time' on things that are not directly relevant to your work because you are curious. But then you are able to, sometimes unconsciously, integrate them back into your work."

At first look Gadiesh doesn't appear to be someone with a diverse occupational background. She joined Bain Consulting in 1977 and has stayed ever since. "I see what you mean," she says when I point this out. "But it is not really a paradox. I have worked in all the fields there are." Her approach to work at Bain has been anything but specialized. She calls herself an expert at being a generalist, or an expert-generalist, a term she has coined within the firm to describe someone who is adept at generating innovative strategies and insights for any industry. She has never chosen a practice within Bain and she has worked in almost all industries. She does not necessarily have to understand how to make steel (although she does) in order to understand aspects about strategy for the steel industry. "I know that when I look at the steel industry, I do so from a perspective that is very different than people that have worked within the industry for decades." Gadiesh believes that such insights can come from fields other than just business. For instance, she reads close to a hundred books a year, and none of them are about business.

Orit Gadiesh has infused these values into the Bain culture and, more than perhaps any other major consulting firm, Bain's organization reflects these values. While Bain certainly has practices and experts, its consultants work in areas outside their specialties. You can find the head of the health care practice working on media strategy. "He will return to the health care practice with more ideas, and will have brought new ones into media strategy," she explains. "Don't get me wrong. At this point we have experts in just about every business. We have people who can talk about consumer products and high-tech in their sleep. We have to. That's the easy part. But we don't let somebody just do that for their entire career, all the time. That was why I said we make people switch areas and fields. It is fundamental at Bain, a core reason for our success. You become better at your area of expertise when you actually take a chance and do something else."

Gadiesh clearly feels that if the consultants at Bain can find their way to the Intersection, they can better help the firm's clients. This was, after all, what enabled her to excel. But individuals who expect to develop intersectional ideas cannot simply hope that their organization will provide them with occupational diversification. They have to control their own fate. By making sure that we gain exposure to different fields during our career, we set ourselves up for more random concept combinations. Frank Herbert, the author of the science fiction book *Dune*, is a good example of someone who used this approach with a vengeance.[6]

During the Second World War, Herbert was an accredited photographer with the U.S. Navy. He later became a reporter and editor at a number of West Coast newspapers and also worked as a TV cameraman, radio news commentator, and even as a speechwriter for California politicians. Herbert also worked as a social and ecological studies consultant in Vietnam and Pakistan and was a lecturer in general and interdisciplinary studies at the University of Washington. It doesn't stop there. He was also an oyster diver, judo instructor, and jungle survival instructor. In 1973 he was a director and photographer for the television show *The Tillers*. Herbert worked and studied in fields as diverse as undersea geology, psychology, navigation, and jungle botany. And, of course, he was a prolific science fiction writer, having published more than twenty-five books by the time he died in 1986.

So maybe Herbert was a bit unusual. Few could ever hope, or would even want, to emulate his incredibly productive and varied life. Yet Herbert is a fascinating case of how occupational diversification leads to the Intersection. Many consider *Dune* the best science fiction book ever written.[7] The book and its sequels combine profound theories about ecology, religion, desert survival techniques, philosophy, and the politics of war into a gripping story. It was Herbert's diverse occupational experiences—and his ability to infuse the resulting knowledge into his story lines—that ultimately led to his literary successes.

In fact, successful innovators tend to work on several interrelated projects at once, rotating within a "network of enterprises" according to whatever appears most promising at the moment. Both Thomas Edison and Charles Darwin, for instance, had many journals and portfolios where they could store notes and articles relating to any number of projects that they were working on. They would regularly review their notes, read over past projects, and reconsider earlier ideas, including the ones that didn't work out. While reviewing their archives with fresh eyes, they might find connections to a current dilemma and perhaps come up with a new solution.[8]

Work with Diverse Groups of People

D URING WORLD WAR II, the Allies were fighting a losing battle against the German navy. When a German submarine spotted an Allied convoy, it would send a coded signal to other German submarines in the area. These submarines would then gather into a group
formation, known as "wolf packs," and attack the ship with punishing
success. The Germans were amazingly effective; between 1940 and
1941 they sank more than fifty ships a month, leading to total casualties
exceeding fifty thousand.

The Allies were helpless against these attacks because they were
unable to break the German coding system, which was produced via a
coding machine known as the Enigma, the most formidable of ci-
phers. British intelligence therefore built the most formidable of code-
breaking groups, headquartered in a large Victorian mansion called
Bletchely Park. Although cryptologists had traditionally come from the
field of linguistics, this group also contained mathematicians, scientists, classicists, chess grand masters, and crossword addicts, all of
whom worked together under supreme secrecy. Together this diverse
team managed to break the Enigma and, as a result, turned the tide of
the naval battle.[9]

There is little doubt that diverse teams, like the one at Bletchely
Park, have a greater chance of coming up with unique ideas. I don't
mean diversity only in terms of disciplines, but also in terms of culture,
ethnicity, geography, age, and gender. Diversity in teams allows different viewpoints, approaches, and frames of mind to emerge. Diversity is
also a proven way to increase the randomness of concept combinations. It is often said that one of the reasons for the United States' un-
paralleled innovation rate is its very diverse population.[10] People who
have experienced the innovative power of diverse teams tend to do
everything they can to encourage them.

Steve Miller is such a person. He is the former CEO and chairman
of Royal Dutch/Shell, the world's fourth-largest company. If you talk to

Miller about innovation for any length of time, it becomes clear that he believes diversity is a critical ingredient.[11] Globalization has made it a necessity for a multinational like Shell. "You begin to find that you get some really neat ideas generated from creating a culture where people of different ethnicities, cultures, backgrounds, [and] countries . . . come together," he says. "Invariably you find that the best ideas come from this mosaic of players working together in a team on a project. They will come up with an answer that is different from what any one of them would have come up with individually."

Working with a diverse group of people, then, is a great way to increase creativity. Even though this may seem like an obvious truth, it is remarkable how seldom we use it. People tend to stick to their own disciplines and domains. They stick to their own ethnicities and cultures. Miller often sees managers who logically understand that a team with people from different backgrounds can be more creative since "you can intellectually work your way through that." But most people have a difficult time going from understanding the logic of such an argument to actually applying it, Miller says. He believes it is easier to do if you have actually seen the power of diverse teams, "because then you really know that it works."

Why are we so hesitant about working in diverse teams? The reason is at least in part a function of human nature. Humans have a tendency to stick with people who are like themselves and avoid those who are different. Psychologists have a name for this tendency. They call it the similar-attraction effect. Donn Byrne of the State University of New York at Albany, a pioneering social psychologist in this area, developed a test to study it.[12] Here is how it worked: A group of college students were asked to indicate their attitudes concerning twenty-six topics ranging from premarital sex, sit-coms, and student and professor needs to legalization of marijuana. The researcher collected the answers, and it seemed like the experiment was over.

Two weeks later the participants were informed that they were now in a new study, one that investigated how well people could predict each other's behavior. The students were given scales that showed another subject's attitudes toward the previously mentioned issues. They were then asked to rate the subject in categories, such as how they felt toward this stranger, if they would like to work with this person, and so on. But there *were* no "other subjects" (which is why this technique is called the bogus-stranger technique). Instead, when the experimenter had prepared the scales, he had invented other subjects with attitudes either similar or dissimilar to the student in the experiment. It turned out that in every instance, the student showed attraction toward the bogus stranger if the underlying attitudes were similar to his or her own. The student liked the bogus stranger better, wanted to work with that person, and evaluated the other person more positively in every way.

What is surprising about these results is not that people are attracted to people who are similar; this is something we know from personal experience. What is surprising is how predictable this effect is. Dr. Byrne found that as the proportion of similar attitudes increases, attraction increases. The effect is so predictable that it can be expressed through a regression equation.

The similar-attraction effect can have a devastating impact on our efforts to create diverse teams. Most people, for instance, think they are pretty decent at interviewing candidates for jobs. Some people even claim that they can tell as soon as a candidate walks through the door whether the person is suitable. "When you've been in the game as long as I have you can spot them straight away," they say. Such talk flies in the face of hundreds of studies that have been conducted since the beginning of the century.[13] These studies show conclusively that the unstructured interview has virtually no validity as a selection tool. Such an interview does not give us enough information to understand the candidate's qualifications. There are many reasons for this problem. People tend to search for commonalities in others. Both the person conducting the interview and the interviewee try to find common ground quickly; if they do, they get a good feeling about each other. The result is that people tend to recruit candidates just like themselves. We do this because we are affected by subjective biases, and in particular by the similar-attraction effect. Even if we want to create an innovative environment with different types of people, we face millions of years of evolution that work against such desires. Yet if we wish to bring ourselves, and our organizations, into the Intersection, we must diversify. How?

Professor Robert Sutton of Stanford University suggests a number of methods to make this happen in his book *Weird Ideas That Work*.[14] One of his first weird ideas is to hire people who make you uncomfortable, even those whom you dislike. If you are thinking about recruiting a candidate because "I like her" or "He's just like one of us," these might actually be reasons *not* to hire the person, assuming the job or team requires creativity. Managers can combat this tendency by monitoring signs that people are hiring too many others like themselves (for instance, check the percentage of graduates from the same school, geographic area, discipline, functional background, former employers, age, race, and gender). Sutton also encourages firms to hire people they don't need, at least not yet. This may sound like strange advice at first, but people are more likely to bring something new to the company if they are not recruited to fill an established role. And if they are motivated and engaged, they will be able to find intersections between their skills and the organization's needs.

Something people often fail to appreciate is that the inverse of these suggestions is also true. That is, if you wish to generate intersectional ideas, you should seek environments where you will work with people who are different from you. Put another way: A sure path to inhibit your own creativity is to seek out environments where people are just like you. If you are drawn to an organization because everyone there sees the world the same way you do (whether that means left-brained, right-brained, artistically, financially, or by any other measure), consider just how this will help you. Chances are you will end up in a team with people who act and think like you. Your team will get along great and it will get a lot of things done. But will it be innovative? Most likely not. Everyone comes to the table with a similar mind-set—and they will leave with the same.[15]

Even when organizations have excellent opportunities to make use of diverse groups of people, they often fail to do so. John Donaghue, the director for Brown's Brain Science Program discussed in chapter 1, considers the open and connected environment at the university to be a key reason for their success. But he's seen others struggle to develop such an atmosphere: "Many times I have visited other labs and noticed that they have another research team just on the other side of the corridor, and I would comment on how great it would be since the two groups could meet and exchange ideas. 'Meet?' they would say, 'I don't even know who works across the hall.' And I would be surprised since I consider that type of interaction crucial for our success."[16]

That said, let me quickly acknowledge that this is not easy advice to heed. There is a reason other than genetics for our tendency to hang out with people who are like us: It makes everything so much *easier*. Hiring people we dislike can lead to trouble—arguments and a negative atmosphere. Simply bringing people together from different disciplines and cultures, with varied thinking styles, different values, and diverse attitudes, is *not* the same as putting together an innovative team. Basic problematic group dynamics will work against you unless the group is managed appropriately.

For starters, it is important to depersonalize conflicts. People should be able to disagree with anyone in the group—but not without a reason. Disagreements can make people feel unfairly targeted if an argument is not specific. It is also important to maintain an open environment where all ideas get a fair hearing.[17] Leaders of teams can, consciously or unconsciously, limit the range of ideas among group members. But at the Intersection, we need as many opportunities for random combinations of ideas as possible. A team of diverse people who feel free to exchange and combine their ideas is exactly what can make that happen.

Go Intersection Hunting

IN ORDER to generate intersectional ideas, we must increase the chances for random combinations to occur. This happens when we diversify occupations, as Frank Herbert and Orit Gadiesh have done. This also happens when we interact with people who have backgrounds, attitudes, and cultures different from our own, as did the cryptologists who cracked the Enigma and the teams at Shell. Both strategies focus on increasing the number of random combinations of different concepts by diversifying. But is it possible to bring this process to the surface whenever we need to?

If increasing random combinations is at the core of generating intersectional ideas, it would make sense to intentionally introduce randomness into our thinking pattern. Such a suggestion may seem strange because we rarely do anything randomly. If you wish to think of a better way to route telecommunications messages, for instance, it would feel odd to explore ideas related to, say, ant-feeding behavior. The subject has no apparent relevance to telecom problems and is therefore left untouched. But is it possible that such an approach, however counterintuitive, could yield significant, practical, innovative ideas?

It is, and you will see how the ant/telecom example plays out later on in this book. Both academic research and a great deal of anecdotal information have clearly shown the advantages of introducing randomness in our thought patterns. I call purposeful efforts to find unusual concept combinations *intersection hunting* and there are, paradoxically, some structured ways to go about it.[18]

Intersection hunting means that you search for connections in unlikely places and then see where those connections lead. When Edgar Allan Poe had to come up with a new plot for his next story, for instance, he would look up two or three words at random in a dictionary and then attempt to tie them together. If he succeeded he would start writing; if he didn't, he would just look up three new words and try again. Michael Michalko, whom I mentioned in the last chapter, describes another way of going intersection hunting, something he calls "taking a thought walk."[19]

If you are working on a specific problem or are just beginning to structure an idea, you can take a thought walk to enhance the chance of random combinations. During a thought walk you might stroll through your office, into the parking lot, or down the street. Pick up, borrow, purchase, or randomly note items during your thought walk (e.g., fishing rod, water cooler, perfume bottle, door hinge, daffodils, etc.). Do not select things that you think are related to the problem or idea because that would be a planned, rather than a random, combination of concepts. Instead, select items with no apparent connection; your job will be to find one.

When you return from your thought walk, write down the characteristics of each word or item you picked up or made note of. The word *painting*, for instance, could include various characteristics: done in different media such as oil, water, computer, or pencil; can be big or small; usually hangs on the wall; often appreciates in value over time; collectors' item; is found in museums; and so forth. Now try to force a connection between these characteristics and the problem you are working on. Some of the ideas generated might give you a unique insight that could solve the problem. Michalko gives the following example of a successful thought walk:

> *A few months back, a group of engineers were looking for ways to safely and efficiently remove ice from power lines during ice storms, but they were stonewalled. They decided to take a "thought walk" around the hotel. One of the engineers came back with a jar of honey he purchased in a gift shop. He suggested putting honey pots on top of each pole. He said this would attract bears. The bears would climb the pole to get the honey, and their climbing would cause the poles to sway and the ice would vibrate off the wires. Working with the principle of vibration, they got the idea of bringing in helicopters to hover over the lines. Their hovering vibrated the ice off the lines.*

Any source of random inspiration is fair game in the hunt for intersections. Take a break from what you are doing, grab a notepad, and start forcing connections between unrelated observations and the problem at hand. With time and luck, you will find a concept that triggers an unusual insight. Before you catch a flight, for instance, buy a couple of magazines you usually do not read, select a page in one of them, and try to connect what is on that page with something you are working on. If you can't find a connection, or if the connection seems way out of line, flip the page, but don't stop on material that has an obvious relevance to the problem. If you are, say, writing a travel guide, look to cookbooks for ideas. Or, the next time you are planning a meal, look in a travel guide for inspiration. Either way, you've just increased your chances of finding an unlikely intersection of fields and the remarkable ideas that follow such a discovery.

Whether we like it or not, the process of innovation is dictated by random combinations of different concepts. Individuals and teams who often break new ground know this and therefore maximize their chances of finding intersectional ideas. They do it by introducing diversity into their occupations, teams, and encounters. It worked for Richard Garfield. Wizards of the Coast has continued to grow the collectible card game industry in a directional fashion. Garfield, on the other hand, has moved on, trying to find the next big thing. "There has been a great deal of bringing out old ideas from the closet and figuring out where to put them in the future," Garfield says, " so I'm looking at a lot of different disciplines, trying to combine them in order to come up with a new game."[20] One that is very different from anything the game world has ever seen.

Preparing for the Explosion

THE STORIES AND STRATEGIES presented so far are intended to help you find that space—the intersection—between different fields. There you will have a greater chance of finding concept combinations that are unlikely enough to be considered revolutionary. But, remarkably, this does not by itself completely explain how the Intersection can create the burst of innovation that I call the Medici Effect. There is another force at play—and it is very strong.

seven

Ignite an Explosion of Ideas

SUBMARINES AND TUBULAR BELLS

I T WAS A CALM EVENING in the summer of 1982. The prolific inventor and engineer Håkan Lans and his wife, Inga, had been sailing through the Stockholm archipelago for the past couple of days, enjoying an unusual spell of beautiful weather. Toward the late afternoon they hooked up to a small island and Lans decided to go for a quiet walk. He climbed to the top of the island and sat down to relax.[1]

Until this sailing excursion, Lans's mind had been occupied with a particularly complex issue. About a year or so earlier he had learned about the U.S. military's new Global Positioning System (GPS), a constellation of satellites deployed to aid armed forces in navigation and position location. Today GPS supports a wide range of commercial uses—from tracking stolen cars to tracking one's own kids—but at the time it was entirely new.

Lans realized, even back then, that the GPS network of satellites could be used differently, as part of a much larger technology that would make airplanes and ships safer to navigate. He envisioned a system where every single airplane could coordinate with all others, rather than relying on the expensive and accident-prone system of radar-manned towers. The system he imagined would save billions of dollars, save lives, and also free up space in increasingly congested airways.

The only problem: It was not possible to execute his vision. Lans faced what seemed like an insurmountable physical limitation. In order to make this idea a reality, all airplanes would have to broadcast their positions to other close aircraft virtually at the same time. The current technology to make that happen, TDMA (Time Division Multiple Access), was woefully inadequate. Perhaps the best way to understand the limits of TDMA is to imagine thousands of people yelling out their position at the same time. It would be impossible to hear what some people said because their voices would be crowded out by other voices, like the early-morning chatter between animals in a rainforest. Hence the system would be useless.

On this island top, so far away from computers and technology, the question of his navigational system suddenly came into focus. Looking over the glittering sea, he had an idea. What if an airplane could broadcast its position only when it was approaching another airplane? After all, wasn't that the only time a collision was possible? Wouldn't that free up some airtime, allowing planes to communicate in a more orderly way? Maybe it would, he thought. Maybe it could. . . .

The way Lans describes the moment, his breathing slowed and the world around him seemed to stop. He started to tremble as one association hooked onto another and an entire vision of related ideas and inventions unrolled before his inner eye. Lans stood up and ran back down to the sailboat. He needed to get back to work.

The Relationship Between Quantity and Quality of Ideas

IS THERE SUCH A THING as a defining characteristic for success-ful innovators? Is there one thing that, more than any other, holds true for people who develop groundbreaking ideas? Actually there is, and it is this: *The most successful innovators produce and realize an incredible number of ideas.*

The strongest correlation for quality of ideas is, in fact, quantity of ideas. A closer look at the number of new products, songs, books, scientific papers, strategy concepts, ideas—any category, anywhere—reveals that they are not evenly distributed. In any given field of creative activity, it is typical to find that around 10 percent of the creators are responsible for 50 percent of all the contributions.[2] Some individuals or creative teams will come up with ten, a hundred, or even a thousand times more ideas than their peers. Not only that, those who have created the most are also the ones who have the most significant innovative impact. This was true in the past; Pablo Picasso, for instance, produced 20,000 pieces of art; Einstein wrote more than 240 papers; Bach wrote a cantata every week; Thomas Edison filed a record 1,039 patents. This holds true today. Prince is said to have over 1,000 songs stored in his secret "vault," and Richard Branson has started 250 companies.[3]

Consider an author like Joyce Carol Oates, one of the usual suspects for the Nobel Prize in literature. She published her first novel in 1964 and, almost four decades later, had published a total of forty-five novels, thirty-nine story collections, eight poetry collections, five dramas, and nine essay collections and contributed to sixteen anthologies.[4] She writes stories the way some of us sign greeting cards. This is the kind of person who innovates.

Why are some innovators so productive? And what, if anything, does that have to do with the Intersection? This chapter will answer both of these questions because they are critical to understanding why the Intersection is so powerful in creating the Medici Effect. The bottom line is this: The intersection of fields, cultures, and disciplines generates combinations of *different* ideas, yes; but it also generates a *massive number* of those combinations. People at the Intersection, then, can pursue more ideas in search of the right ones.

Virtually every person or team I've met while learning about the intersection emphasized the need to try many ideas in order to generate something groundbreaking. Perhaps no one exemplifies this better than Håkan Lans. Although you've probably never heard of him, he is one of the most prolific and successful innovators of our time.

The Producer of Ideas

THE FIRST THING that struck me about Lans was an appearance of modesty.[5] He lives unassumingly, well below his means, in a nice but not extravagant house, and he drives a nice but not extravagant car. He doesn't seek out the limelight, but he is clearly not shy. Once he gets going, Lans can talk for hours about virtually anything.

The second thing that struck me was that Lans is different from most people described in this book in that he grew up and lived mostly in one place his entire life, around the city of Stockholm. Lans has broken down his associative barriers by *learning differently*; he is self-taught in virtually every discipline of technology and engineering. Lans is also adept at finding intersections between many of those fields, this, he will tell you, is the reason for his success. Today he is one of the most well regarded scientists in Sweden, even without a formal Ph.D. What is his secret? How did he end up here?

Modest living aside, his life has some of the trappings of a good spy novel, including international espionage, high-stakes courtroom battles, and patent thefts. He single-handedly took companies such as Hitachi to task for copyright infringements and challenged world bodies such as the United Nations and the European Union. But he also produced ideas and innovations at a prolific pace.

His most significant innovation is probably the development of the navigation system called STDMA (Self-organizing Time Division Multiple Access). That flash-in-the-sky insight Lans had on the island ultimately launched an incredibly ambitious project that took him many years to complete, entirely on his own. Today the system is becoming the world standard for air and sea traffic navigation. It may seem like a once-in-a-lifetime idea, but it is just one example of Lans's continual search for intersections involving various types of vehicles and computing or engineering technologies.[6]

One of his very first childhood experiences as an innovator falls into this category. "It was spring, and all the kids were building boxcars, trying to get them done as quickly as possible," he says. But Lans was always a little different. "I tried to integrate a motorcycle engine into the boxcar. But it was difficult; it took time. The others completed their boxcars and started racing, and they teased me for the fact that I did not have one."

Then, one morning, Lans finally managed to make the boxcar and engine combination work. He started the car and drove to school, stopping by the entrance and letting the engine rev. Soon every kid in school was standing in a ring around him, staring. Lans says, "I just turned the engine off, stood up, excused myself and walked into class. It felt very good, and I think that particular feeling of pride was instrumental in giving me confidence that I could succeed [as an innovator]."

Young Lans also built rockets that flew and exploded. He once blew out the entire kitchen in his home. A couple of years later, when he was seventeen, he decided to build a submarine. He had no money, of course, but managed to patch together a network of sponsors. Lans secured a steel sheet from one firm, got someone else to bend it just so, persuaded a third party to attach a glass, and so on. All according to his designs. He interviewed physicians about how humans breathe and then built an entire life-support system for the submarine. Once the "Yellow Submarine," as he called it, was completed, he took it below water for thirty to sixty minutes at a time. And he went deep. Lans, who had just turned eighteen, managed to bring his little homemade vessel 330 feet below the surface. The Swedish navy had only five submarines at the time.

Lans always exhibited an incredible ability to combine different ideas from different fields. When color television hit the market in Sweden, Lans realized that computer monitors would ultimately display color as well. No one, of course, knew what he was talking about; few even knew what a computer was. But his vision led him to develop a color graphic chip that became the standard for producing color graphics in computers during the 1980s. It is hard to overestimate the impact of this achievement. His chip technology was shipped with basically every computer with a color monitor sold in the world at the time.

There are scores of other examples. In his spare time Lans developed a redesigned airplane cockpit. "It looks like a watch store," he said, referring to the instrument panel the aircraft pilots had to use, and so he pulled together all the essential information into one easy-to-view screen. Early in his career he needed an expensive electronic drawing board for his work, but could not afford one. Instead of buying the big, clunky piece of equipment, he decided to create a smaller, more efficient drawing pen linked to the computer. His invention became, in effect, the first mouse with which one could draw curves, a major improvement over Douglas Engelbert's original mouse (which he later sold to Texas Instruments). Over the years Lans developed computers, underwater acoustic transmitters, cryptography modems, pulse generators—the list seems endless. He built his own airplane just for fun. Later this became a test plane for his revolutionary navigational system.

Lans has never seen himself as an ordinary researcher. "I take the puzzle pieces that basic scientists discover and put them together," he says. This puzzle generates a multitude of ideas. Lans then chooses the opportunities that he believes have the best chance to succeed and tries to make them happen.

By no stretch of the imagination is Håkan Lans typical. He has developed and successfully introduced several world-changing inventions, and he has an unrivaled obsession for combining diverse technologies to produce novel applications. But Lans has something in common with every single person or team who innovates at the Intersection. He produces an incredible number of ideas, and he relentlessly pursues the best of them. And that is his secret.

Why Innovators Are Productive

T HE TRADITIONAL WAY to explain why successful innovators produce a lot of ideas is that they get caught in a "virtuous cycle" where past success breeds future opportunities and success.[7] For instance, if a team of entrepreneurs has been successful with an intersectional idea once, investors will be more eager to fund their next venture. The same argument would then hold true for scientists and artists. A successful researcher who has written an exceptional Ph.D. thesis might get recruited to a prestigious institution with good mentors and a strong network to fund his or her research. All of this leads to a reinforced cycle with an ever-higher output of papers and ideas. This explanation makes sense. And it may very well hold true for directional innovation. But it ignores two fundamental truths about intersectional innovation.

First, it does not take into account that the creative process is random. The random process would suggest that it is not always past success that sets someone up for future innovative success, but rather that both past and future innovative success is more a matter of chance than anything else.

Second, this explanation ignores the fact that groundbreaking innovators also produce a heap of ideas that never amount to anything. We play only about 35 percent of Mozart's, Bach's, or Beethoven's compositions today; we view only a fraction of Picasso's works; and most of Einstein's papers were not referenced by anyone.[8] Many of the world's celebrated writers have also produced horrible books, innovative movie directors have made truly uncreative duds, megasuccessful entrepreneurs have disappointed investors, and pioneering scientists have published papers with no impact whatsoever on their colleagues. Consider Charles Darwin. After having proposed the groundbreaking theory of evolution, he developed the dead-wrong theory of pangenesis, which suggested that acquired traits, such as stronger muscles, could be passed on to offspring. Or look at Sabeer Bathia. He founded the e-mail service Hotmail, which became successful because of a novel marketing device—a sign-up link sent automatically with each e-mail. His next venture, Arzoo, an online service market, incorporated what Bathia felt were several innovative ideas—but the company languished.[9] Clearly, one great innovation does not guarantee another.

So what is going on? Why are successful innovators such massive producers? In his influential book *Origins of Genius*, psychologist Dean Simonton from the University of California–Davis explains why we see this relationship between production and success. He says innovators don't produce because they are successful, but that they are successful because they produce. Quantity of ideas leads to quality of ideas.

There is a certain logic to this argument, partially based on the random nature of creativity. Since intersectional ideas are the result of random combinations of concepts, it follows that the more random combinations one has, the better the chances of coming up with something truly exceptional. That's all well and good, but Simonton went beyond merely logical arguments. He wanted to see if the theory held up to scrutiny, if it actually described what happens in the real world.

Simonton focused his studies on the relationship between the quality and quantity of the creative output from scientists. When a scientist publishes a paper, the most reliable way to measure the quality of that paper is by how many other scientists have referred to it. If a lot of other scientists refer to a particular paper, it is likely that it had a notable impact, maybe even launching a new field. The vast majority of scientific papers receive very few citations, as the referrals are called, while a few papers, the breakthrough ones, receive hordes of citations.

Simonton verified that the relationship between quantity and quality indeed holds true. The number of papers a scientist publishes, for instance, is correlated with the number of citations the scientist receives for his or her top three works. In other words, the best way to see who has written groundbreaking papers is to look at who has published the most. You can test this a hundred different ways, but the results come out the same. The length of a bibliography of a scientist in the nineteenth century predicts how famous that person is today. The best predictor for who will receive distinguished honors, such as a Nobel Prize, is the number of publications the person has published. In fact, the best predictor for having a grant proposal approved is the total number of grant proposals written.[10]

Simonton then did something quite intriguing. He looked at individual scientists' careers. If the virtuous cycle theory were true, you would expect to see an increase in quality of papers after a successful one was published. But you don't. Scientists produced breakthrough papers at random points throughout their careers, but they had the best chance of writing them when they published a lot of papers. The best predictor for when scientists produce their best works, their most exceptional contributions, is actually when they produce the most. Incidentally, this was also when they had the greatest chance of writing their worst papers, which is what you would expect given the random nature of creativity.

Simonton also found that this relationship holds true for artists. Classical composers, for instance, produced most of their masterpieces during the same period when they produced most of their failures. Just because someone has developed a groundbreaking idea once does not necessarily mean that he or she has a better chance of doing it again. Instead, the best way to beat the odds is to continually produce ideas. This is why innovators are so productive.

The Explosion at the Intersection

THE MOST FASCINATING IMPLICATION of Simonton's research, however, is how beautifully it explains the Medici Effect at the Intersection. Why is the intersection of disciplines or cultures such a vibrant place for creativity? We discussed one reason in the last two chapters: It increases the chances that an idea will be good because it brings together very different concepts from very different fields, as in the case of the game Magic. But there is another, stronger, reason for its power. When you connect two separate fields, you also set off an exponential increase of unique concept combinations, a veritable explosion of ideas. Or, to put it succinctly, if being productive is the best strategy to innovate, then the Intersection is the best place to innovate. The following story will show you why.

Richard Branson, founder of successful Virgin Group, got his lucky break in 1971. Given the force of his personality, there is little doubt he would have made Virgin happen one way or another. But as we've just learned, you need that lucky break. And Branson got his when he met the shy hippie teenager Mike Oldfield. It turned out that Oldfield had some strange new ideas about music, and Branson wanted to start a record label. When they struck a partnership, the teenager went on to become one of Great Britain's most successful musicians, and Branson went on to become one of Great Britain's most successful entrepreneurs. The album that catapulted both of their careers was called *Tubular Bells*.[11]

When the album was first released, sales were low because Branson had no money to promote it. But that changed as word of mouth started spreading. About a year after its release, *Tubular Bells* had climbed to the top of the U.K. charts. It held that spot for an incredible fifteen straight months. Today it has sold about 16 million copies worldwide and *still* sells around 100,000 copies a year.[12]

This feat seems even more spectacular when you consider that *Tubular Bells* was unlike any other album preceding it. It was a strange mixture of rock and classical music. The combination of these fields was deep; this was definitely not a rock band playing classical tunes, or a symphony playing pop songs. No, *Tubular Bells* sat right at the intersection of the two fields, combining elements that could be found in both domains. But what, exactly, happens at such an intersection?

Say that you are a rock musician around 1973, when Oldfield released *Tubular Bells*, and say that you are trying to come up with a new sort of music. One way to approach this challenge would be to break down the components that actually constitute a rock song and look at different ways of combining them. Although there are many variations and concepts, for the sake of this example, let's look at three major groups of concepts that define rock music: instruments, structure, and vocals.

> *Instruments*: Rock in the early days was quite a rigid music form in terms of the instruments used. Bands usually consisted of guitar, drums, and bass. Occasionally other instruments were included, such as the saxophone and the piano, but the stereotypical band was pretty simple. Let's say that the average rock composer used four instrument combinations.
>
> *Structures*: Rock music was also fairly limited in its structure. The number of chords used in rock songs tended to be quite low. Moreover, almost every song consisted of two or three verses with a chorus in between. Let's say a rock musician could choose from twelve different structures.
>
> *Vocals*: In contrast, rock employed a variety of voice concepts. Voices could be hushed, raspy, strong, weak, smooth, soulful, and so on. It was not even necessary for people to know how to sing to be considered rock musicians. Bob Dylan had no clue, but that did not stop him from becoming one of the greatest artists ever. Let's say a rock musician had fifty voice concepts to work with.

How many combinations, then, could the average rock musician generate based on these variations? How many times could he combine different instruments with different structures and different vocals before he ran out of combinations? By simply multiplying the variations in each group, we see that a rock musician in this example has 4 × 12 × 50, or 2,400, combinations to work with when developing new music. The musician wouldn't necessarily actively try to combine these areas of music (although this can be a good idea when you go intersection hunting), but they are subconsciously part of the process for generating new music ideas.

Let's switch gears now and look at classical composers. They have a very different set of choices available to them.

> *Instruments:* Classical composers can choose from a wide range of instruments. Symphonies, for instance, can include violins, horns, flutes, harps, gong-gongs, and drums, among many, many others. Let's say a classical composer has thirty instruments to choose from.
>
> *Structures:* Classical composers allow themselves much more variation in the number of structures than most rock musicians do. Music tends to flow and not rely on repeated sequences. Pieces can also vary greatly in length, with some pieces longer than thirty minutes. Let's say for the sake of this example that a composer of classical music can choose from some forty structures.
>
> *Vocals:* Classical music has few vocals. Strictly speaking, these are simply not included in a symphony. In other compositions they tend to be in the form of a choir. Let's say that a classical composer has two choices.

If we calculate the variations as we did for the rock musician, we find that a classical composer can choose from a total of $30 \times 40 \times 2$, or 2,400 concept combinations when trying to come up with new music. The actual number is of course higher, but the big strokes of this example remain faithful to the differences between rock music and classical music.

Now we get to the key point of this exercise. If a person has knowledge of both rock and classical music but views them as separate fields, he can choose from 2,400 combinations in either genre when looking for new musical ideas. But what happens if this person has been able to break down the associative barriers between the two fields? What happens if this person steps into the intersection of the two fields, the way Mike Oldfield did with *Tubular Bells*? It would seem that the number of available concept combinations goes up dramatically since it is now possible to freely mix and match ideas between the domains. And it does. In fact, the number rises exponentially (see figure 7-1). Such a

FIGURE 7-1

**The Medici Effect: An Exponential Increase
in Concept Combinations**

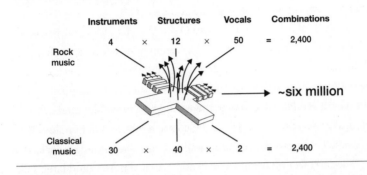

person has 2,400 × 2,400 concept combinations available. That is equal to almost *six million* new ideas—5,760,000, to be exact.

If this number seems staggeringly high, that's because it is. This is what I mean when I talk about the power of the Intersection. This is the heart of the Medici Effect. By breaking down associative barriers and stepping into the intersection between fields, the number of available idea combinations increases beyond anything we can achieve in a single area.

This, then, explains why diverse teams can be more creative than homogeneous groups. It explains why diversifying occupations can increase our output of exceptional ideas. The intersection of fields not only provides the perfect environment for widely different ideas to come together, it also makes it possible for lots of different ideas to do so.

Living with the Explosion

MIKE OLDFIELD lives and breathes at the Intersection, which explains his inexhaustible output of new, interesting music. The guitar was and has remained his core instrument, but Oldfield played more than twenty instruments in *Tubular Bells*. He used vocals sparingly, except for a section called the "piltman song," which Oldfield recorded after having swigged half a bottle of whiskey. Vocals grew increasingly important in later albums, though.

David Bedford, a friend of Mike Oldfield's who ultimately rescored *Tubular Bells* for symphony orchestra, commented a few years after the album had been released: "He stands out in the rock scene because he's the only one who uses a sort of logical construction to his pieces, and they have a semiclassical feel to them. And he'd probably stand out in a classical concert situation in that he'd have a rock feel to him, because his whole background is rock and so that tinges everything he does."[13]

The intersection of rock music and classical music (and later, folk music and electronic music) has provided Oldfield with more combinations than he can use in an entire lifetime. Just as Marcus Samuelsson's food creations sometimes defy expectation, Mike Oldfield's combinations may seem impossible. For instance, in a segment from *Ommadawn*, generally considered one of his best albums, he plays an electric bouzouki, the bagpipe, and the guitar. In another section of that recording he overdubs an electric guitar sixty-four times. This has the same effect as having sixty-four guitarists simultaneously playing the same piece of music, and is reminiscent of how a classical composer would approach the same section. The combinations work, and they work well.

The explosion of ideas at the Intersection, then, is what makes it *possible* for innovators to produce so many remarkable ideas. It gives them an incredible advantage. Oldfield, for instance, has kept his pace and had released more than twenty-five albums by the turn of the millennium, with no signs of stopping. Some of them failed spectacularly, others sold millions.[14] All of them were part of the explosion.

eight

How to Capture the Explosion

MACGYVER AND BOILING POTATOES

LINUS PAULING, Nobel laureate in both chemistry and peace, once said, "The best way to get a good idea is to have a lot of ideas." As the previous chapter illustrates, the explosion of concept combinations at the Intersection unlocks a massive number of potentially groundbreaking ideas. What you have to do now is capture them. This, however, is not an automatic process. Just because you *can* potentially access all of these ideas does not mean that you *do* access them. How, then, can you seize the myriad opportunities at the Intersection? There are at least three ways to proceed:

➤ Strike a balance between depth and breadth
➤ Actively generate many ideas
➤ Allow time for evaluation

Strike a Balance Between Depth and Breadth

THE INTERESTING THING about the mathematics of the Intersection is that even if you knew only a fraction of the concepts
in either rock or classical music (to take the example used in the previous chapter), you would still be ahead of the game. If you multiply
2,400 by 600, for instance, you get 1,440,000 combinations, quite a re-
spectable number. Although encouraging, this also seems a bit strange.
If we push the explosion idea to its extreme, it would appear to be far
better to know a tiny bit about hundreds of fields than a lot about just
one. For example, if you knew, say, a hundred concepts from fifty fields
and had the ability to associate freely among all of them, you could the-
oretically access more concept combinations than there are atoms in
the universe. No one is that innovative, not even Håkan Lans.

The reason the world does not work this way is that we must strike a
balance between depth and breadth of knowledge in order to maximize our
creative potential. Too much expertise, as we have seen, can fortify the
associative barriers between fields. At the same time, expertise is clearly
needed in order to develop new ideas to begin with. It would be unwise to
attempt to change the field of rock if you could not even strike up a tune,
and it would be difficult indeed for a biotechnology company to innovate
drug development without knowing quite a bit about life sciences. Just
how much expertise, then, is required to ignite the perfect explosion?

One way to handle the need for broad yet deep knowledge is to
team up with someone who has a different knowledge base from yours.
As we saw in chapter 6, teams with members from different fields are
more likely to find intersections, assuming they can break down the
barriers between fields. This may, in fact, be the most common approach for generating new ideas. But how does it work for individuals?
Where is this knowledge balance for someone like Mike Oldfield? Just
how well does one need to understand the concepts of a particular field
in order to effectively combine it with another?

The people I have met provide some clues. Most gained knowledge in one specific area before striking out to other fields. I am not talking about world-leading expertise here, but enough to call it a core competence. Mijail Serruya at Brown's Brain Science Program says that no matter how broadly others view him, he can "at least teach a second-year course in neurology." Orit Gadiesh emphasizes that although consultants at Bain can switch practices, most still maintain an area of expertise. Mike Oldfield's love of the guitar shines through in virtually every one of his albums, and Marcus Samuelsson started his career cooking traditional Swedish food. Although it may not be absolutely necessary to approach intersectional innovation by initially concentrating on one area, it can be very helpful. Here's why.

The person who understands many fields and is able to break down the barriers between them all would indeed have access to an incredible number of concept combinations. But such a person faces one big problem. That person would have a much tougher time understanding just *how* to make an intersectional idea happen, or if it even *could* happen. It is one thing to say that one can combine rock music with classical music. It is quite another to actually pull it off.

Actively Generate Many Ideas

ONCE YOU STEP into the Intersection, you need to grab as many unusual ideas as possible. Unfortunately, that behavior does not come instinctively. Consider the following exercise:

A brick manufacturer is experiencing a sharp decrease in sales. The manufacturer is looking for different uses for brick to improve its marketing efforts. You are called in to help. Take some time to think about this problem and write down all of the solutions that come to mind.

How did it go? If you are like most people, you wrote down three to six solutions, such as using brick for a wall, house, chimney, or walkway. Quite possibly you had some problems breaking out of the traditional uses of bricks. You may even have had a couple of ideas that you didn't write down because you didn't consider them very useful and so waited until a "really good one" popped up. This exercise is taken from the book *Conceptual Blockbusting* by James Adams, and it highlights a very common dilemma in trying to think of alternative solutions to a problem: our inherent hesitation to generate multiple ideas.[1]

The funny thing is that we often take a "batch" approach to certain tasks in life. When we boil potatoes, we peel and then cook all of them at the same time. We don't peel and cook them one by one because that obviously would be a complete waste of time and energy. But we often develop ideas this way. If we get an idea that seems promising, we tend to delve deeper into the idea until it either works or it doesn't. If it isn't successful, we start over with another good idea. But this is not the best way to use our time or creative energy. In order to maximize the power of the Intersection, we should generate many ideas before evaluating any one of them. Take a couple of minutes to consider the second part of this exercise:

Take a blank piece of paper and list at least thirty uses for bricks.

What happened this time? You probably listed far more possibilities for the use of brick than in the first part of the exercise. Compare your recent list with the earlier one. Does the second list contain interesting ideas that the first one lacks? One of the best ways to brainstorm privately is to place a target for the number of ideas that you wish to generate before you start considering whether they are any good. The goal is to force you to think far beyond the usual ideas that come to mind. J. P. Guilford, who conducted some of the association tests I discussed in chapter 3, has proven that the first ideas you think of are the common ones, the noncreative ones, like using bricks to build a wall.[2]

The last ideas you think of, however, tend to be more creative. At this point a brick can become a table leg or boat's ballast.

When you're trying to generate better ideas, even to solve a fairly simple problem, sit down and work through a real brainstorming session. This can be done not just at the beginning of a project, but at any time you need some fresh thinking. To innovate, after all, you must test lots of ideas. Mike Oldfield, for instance, did 2,300 recordings of *Tubular Bells*.[3] Thomas Edison conducted more than 9,000 experiments to develop the light bulb and over 50,000 experiments to develop the storage fuel cell. Edison, in fact, placed a quota on himself for generating new ideas. He needed to think of one minor invention every ten days and a major invention every six months.[4]

When you are done with the process, look over your ideas and evaluate them yourself or together with others; then go to work on those that seem promising. Then save the list. You may wish to return to it since many of the ideas could be useful in the future.

The Issues Around Brainstorming

As we know, brainstorming is one of the most common tools for generating ideas. Tom Kelley, former manager and brother to the founder of IDEO, a San Jose–based design shop well known for its innovations, considers brainstorming essential. The firm envisioned and created the Apple mouse, Polaroid's I-Zone instant camera, the self-sealing water bottle, the Palm V, and many other breakthrough products and services. In his book, *The Art of Innovation*, Kelley considers brainstorming one of the most critical components of IDEO's success: "Brainstorming is the idea engine of IDEO's culture. It's an opportunity for teams to 'blue sky' ideas early in a project or to solve tricky problems that cropped up later on. . . . The buzz of a good brainstormer can infect a team with optimism and a sense of opportunity that can carry it through the darkest and most pressure-tinged stages of a project."[5]

None of this should really surprise anyone. Brainstorming is the most widely used method for a group to generate a large number of ideas on any topic. In his influential 1957 book, *Applied Imagination*, Alex Osborn suggested brainstorming as a method for groups that were solving problems.[6] According to Osborn, brainstorming would greatly increase the quantity and quality of ideas generated by the group. The rules for brainstorming were easy. The group should:

1. Produce as many ideas as possible
2. Produce ideas as wild as possible
3. Build upon each other's ideas
4. Avoid passing judgment on ideas

Brainstorming has since been used in nearly all of the world's largest companies, nonprofits, and government organizations. And the reasons seem obvious. Osborn wrote, referring to brainstorming, "The average person can think of twice as many ideas when working with a group than when working alone." With such odds, it is no wonder that it would be spreading wide and far. But is it true?

The first study to test Osborn's claim came in 1958, only one year after his book had been published. Psychologists let groups of four people brainstorm about the practical benefits or difficulties that would arise if everyone had an extra thumb on each hand after next year. These groups were called "real groups" since they actually brainstormed together. Next, the researchers let "virtual groups" of four people generate ideas around the "thumb problem," but they had to brainstorm individually, in separate rooms. The researchers combined the answers they received from each individual and eliminated redundancy by counting ideas that had been suggested multiple times only once. They then compared the performance between real groups and virtual groups.

The results were not what you would expect. To their surprise, the researchers found that virtual groups, where people brainstormed individually, generated nearly twice as many ideas as the real groups.

This result, it turns out, was not an anomaly. In a famous 1987 study, researchers Michael Diehl and Wolfgang Stroebe from Tubingen University in Germany concluded that brainstorming groups have *never* outperformed virtual groups.[7] Of the twenty-five reported experiments by psychologists all over the world, real groups have never once been shown to be more productive than virtual groups. In fact, real groups that engage in brainstorming *consistently* generate about *half* the number of ideas they would have produced if the group's individuals had pondered the problem on their own. In addition, in the studies where the quality of ideas was measured, researchers found that the total number of good ideas was much higher in virtual groups than in real groups.

These results are confounding. We are used to thinking that brainstorming will enhance a team's creative abilities; this is, after all, why we do it. In general, however, research insists that brainstorming is difficult to get right. Tom Kelley also suggests that there may be more to brainstorming than simply following the original four rules. "The problem with brainstorming is that everyone thinks they are doing it," he says. "Brainstorming is practically a religion at IDEO, one we practice nearly every day. Though brainstorms themselves are often playful, brainstorming as a tool—as a *skill*—is taken quite seriously."

Diehl and Stroebe set out to understand why brainstorming was such an unpredictable methodology. They arranged three experiments to test three separate theories in an attempt to isolate the most crucial factor for such a counterintuitive effect. The first theory referred to the "free rider phenomenon," where some participants of a group would essentially relax and rely on others to come up with new ideas, since ultimately the contributions would be anonymous. The second theory was "evaluation apprehension," which suggested that some group members avoided expressing wild or original ideas based on how the other members of the group would privately judge them. Both of these effects seemed to play some role, but they were not very significant. Instead, it was a phenomenon called "blocking" that was responsible for the vast difference between brainstorming in a group and doing so individually.

In a brainstorming group only one person can speak at a time, although not necessarily in any particular order. If everyone spoke at once, no one would hear what the others said. But this presents a big problem for us humans. Our short-term memory is not capable of developing new ideas and at the same time keeping the old ones in active storage. If we become blocked in reporting our ideas because we have to wait for someone else to describe theirs, we may forget them altogether. This makes a big difference in our output since we cannot simply call out an idea when we think of it; we have to wait until the current speaker has finished. And when we do get a chance to describe an idea, we may get to offer only one or two comments before someone else breaks in. This explanation also supports the general finding that the larger the brainstorming group, the *fewer* the ideas produced compared to the number generated by a virtual group of the same size.

Fixing Brainstorming

So should we all stop brainstorming? No, I don't think so. *Done right*, brainstorming is a highly effective way to actively generate intersectional ideas. Research results suggest that some small, but very significant, changes to common brainstorming greatly enhances effectiveness.

First, before the group meets, schedule fifteen to twenty minutes for members to brainstorm individually. Then they do not have to worry about forgetting their original ideas when the group phase of the brainstorming begins. This also forces the facilitator to develop a well-formulated problem statement, which has been shown to make brainstorming more effective. Second, bring the members together and start a group session. Don't let people just take turns reading down their list. (It will stifle the momentum and make it difficult for people to actively build off each other's ideas.) Instead, keep everyone involved, and keep the pace and action high. By the time you're finished, the combined ideas from all individuals should be on the board, and most of them should have been discussed.[8]

Diehl and Stroebe's research results suggest yet another way to sidestep the problems with traditional brainstorming—a technique called brainwriting. While brainwriting, people simultaneously generate written ideas on the same problem, building off each other's ideas without speaking at all. Here is how you do it:[9] Everyone sits at a table together, each person with a blank sheet of paper. Another blank sheet is in the middle of the table within everyone's reach. The basic problem to be solved or explored has been clearly described or written down. At the start of the session, each person writes (or sketches) one idea on the sheet in front of them, tosses that sheet into the center of the table, and then picks up a sheet put in by someone else. The person reads the idea on that sheet and tries to build on it in some way. Whether or not they can directly build on it, they write another idea, toss the sheet into the center, and continue. Whenever anyone picks up a sheet from the center of the table, they read through prior ideas, trying to make connections and ignite sparks of new ideas. This approach could also be used successfully in an online virtual environment where people continuously comment and build off one another's ideas.

Allow Time for Evaluation

THERE IS AN EPISODE of the popular 1980s TV show *MacGyver* that goes something like this: The hero, MacGyver, has to save two scientists trapped in a high-security underground laboratory while an acid leak threatens the entire New Mexico water system. Although he is very pressed for time and resources, he solves the problems with remarkable ingenuity. For instance, in order to lift a steel beam, he ties a knot in a fire hose and builds water pressure strong enough to push the steel beam out of the way. A couple of minutes later, he manages to stop the acid leak by plugging the cracks with milk chocolate bars. (Honest.)

Actually, all *MacGyver* episodes featured similar last-minute—or last-second—creative challenges.[10] And they served to illustrate a common belief among executives and others that we generate our best ideas when time is tight and deadlines are looming. We supposedly do our most creative work while high on adrenaline and caffeine but low on resources—time in particular. But is *MacGyver* representative of the real world?

In one of the most comprehensive and ambitious attempts ever at understanding creativity in action, Harvard Business School professor and leading creativity researcher Teresa Amabile showed that this perception is a myth. In the study, Amabile and her colleagues followed 177 employees in twenty-two project teams from seven companies for the entire duration of a project, in some cases as long as six months.[11] These teams were not just any teams; they were considered the "creative lifeblood" of their organizations. The researchers e-mailed all team participants a daily questionnaire asking them about their project and how they felt about it. With over nine thousand responses, they could then search the data for trends.

What they found was fascinating. Not only did they find that people are *less* creative under serious time pressure, but people actually believe that they are *more* creative during these times. In addition, they found that creativity decreased not just on the day of intense time stress, but also on the following day, the day after that, *and the day after that.*

In a few instances time pressure did inspire creativity for some people. Specifically, the person had to be fully focused on the project at hand, not distracted by meetings or memos, and working with just one or two other collaborators; also, the time pressures had to be real. Situations like this, however, were exceedingly rare in the companies they studied. Sometimes the teams were placed under artificial time limits, but this often backfired. Amabile writes, "management perennially put teams under severe and seemingly arbitrary time and resource constraints. At first, many team members were energized by the fire-fighting atmosphere. They threw themselves into their work and rallied. But after a few months, their verve had diminished . . . because pressures had proven meaningless."[12]

In fact, if you want to capture intersectional ideas, your best bet may be to take your time. There are at least two reasons for this. First, it is critical to postpone judgment of new ideas. Our minds will quickly judge the value of an intersectional insight by comparing it to what is known to work within an established field. But these fields are not good guides for evaluating ideas that result from random and unusual concept combinations. Instead, intersectional ideas must be evaluated from a different perspective, one that does not come instinctively. You are therefore better off waiting to judge your insights when you have some time to think them through.

Consider Håkan Lans. He considered a more general idea for more than a year before he had the revolutionary insight that led to his navigational system. What if he had been under intense time pressure? Would he even have taken the break to go sailing with his wife? Remember Richard Garfield, the designer of Magic? It took him eight years to gain the sudden insight of combining games and collectible items—and at first he didn't even know what that insight meant. Garfield, as a trained game designer, could easily have brushed off such a "moment of truth" as silly and moved on, tinkering with various aspects of traditional game design. But he didn't. Instead he played around with the notion. "It was not until a month or two later that I pulled out this card game I had been working on for a while . . . and I realized maybe I had the roots of . . . a game there," he says.[13]

Taking time to judge unique insights may sound like very simple advice, but it is actually quite difficult to execute. Our mind tends to sort through ideas quickly, and unless we use some type of recording system, it will kindly get rid of those thoughts it deems unworthy. Many ideas have probably passed through your mind as you have read this book, but how many of them do you remember?

Probably the best insurance against prejudging ideas is to write them down or diagram them when they occur to you. This will allow you to return to the idea at frequent intervals. Then, if an idea suddenly seems more attractive, you can examine it more closely. Keep a notepad by your bed, a small memo pad next to the shower, and a bound notebook with you at all times. Taking notes in the car is a bit more problematic. Nevertheless, some of the best ideas strike us while we are driving alone. Try using a tape recorder. More important than keeping notebooks handy is actually using them. Getting used to recording ideas, thoughts, and insights requires commitment. Once you develop this habit, though, you will wonder how you ever made it through the day without it.

Taking time to evaluate new ideas is important for another reason. In chapter 5 I talked about the incubation period and how it leads to flash-in-the-sky discoveries. The incubation period is the time between when one stops thinking heavily about a problem and when one suddenly, subconsciously, comes up with a solution. The incubation period is so well documented in creativity research that it is simply bad planning not to include time for it while working on a project. It may very well be that we work harder and are more focused under a tight deadline, but how often have you completed a project, an assignment, or anything requiring some level of creativity, only to get a better idea once you were done? That is, after the deadline. The incubation period suggests that we should work in a very different way. It suggests that we should start by working hard and in a focused manner on a problem or idea and develop it as far as possible. Then we should wait, move on to something else, and forget about the problem for awhile. When we return to the project a few days or weeks later, other ideas, usually more original ones, will have presented themselves.

From Ideas to Innovation

So far we've been concerned primarily with ideas. We have looked at why someone like Marcus Samuelsson has a relatively easy time breaking down associative barriers between unconnected fields, and how we can do the same. We have studied in detail how the revolutionary game Magic was created through a random combination of existing concepts, and how we can engineer such clashes. Finally, the preceding two chapters discussed why innovative people like Håkan Lans are productive people, why the Intersection is the best place for generating groundbreaking ideas, and how we can capture them. The question that follows is, what happens next? Once we've discovered these fascinating ideas, what do we do?

Well, we have to execute—to realize those ideas. Otherwise we will never innovate. Håkan Lans told me the following story that, I believe, perfectly illustrates what happens when, for a variety of reasons, we fail to turn ideas into action.

During the years I have been contacted by a large number of people that wish to tell me about an idea of theirs. I'm thinking specifically about one person. He is well educated and has a Ph.D. He usually calls me every couple of years to talk about an amazing new idea he has, and usually it really is a great idea. At the same time he laments the stupidity of the world that simply could not see how bright the idea is. But he has never, ever, tried to make any of his ideas happen. Well, he called me a couple of years ago and presented one of those super-brilliant ideas and once again started his complaining about how the world ignored his insights. He said that he didn't need that much money, maybe just about $100,000, to make it happen. And he asked if I knew anyone that could help him out with funding. I usually don't meddle in other people's projects, but this time I made an exception. I called a couple of people and they were very positive and told me they would meet with him.

Six months later this person calls me again and tells me about another idea he has. I was a bit taken aback and interrupted him. "But . . hold on, what happened? They never got in touch with you?"

"Oh, yes, they got in touch with me," he answered.
"Oh . . . so you didn't get the money?"
"No, no," he said, "I could get virtually as much as I wanted."
"So, what was it then?"
"Well, you see, this new idea is so much better."[14]

The explosion of concept combinations at the Intersection can offer a myriad of uniquely combined, extraordinary ideas. Coming up with great ideas, however, does not guarantee an innovation. You must make those ideas happen.

PART THREE

Making Intersectional Ideas Happen

nine

Execute Past Your Failures

VIOLENCE AND SCHOOL CURRICULA

IT WAS AROUND THREE O'CLOCK in the morning on a January night in 1978. A young man, not yet in his twenties, had just walked into the emergency room at Brigham and Women's Hospital in Boston. He was tightly pressing a bloodstained shirt against a deep cut above his eye. Deborah Prothrow-Stith was only a third-year medical student at the time, midway through her surgical rotation, and her task this particular night was to practice stitching up patients. While she took care of the man, he told her what had happened. He'd been at a party and some guy he barely knew had offended him. One thing quickly led to another and suddenly they were squaring off amidst a ring of onlookers. Seconds later a knife flashed across his face. An inch lower and his eye would have been history. Instead, it was gleaming with anger. When Prothrow-Stith was done he turned to her and spoke words she would never forget: "Look, don't go to sleep because the guy who did this to me is going to be in here in about an hour and you're going to get all the practice stitching you need!"

Then he left.

That experience was an epiphany for Prothrow-Stith. It led to an insight that propelled her right into the intersection of two completely separate fields—violence prevention and health care. What is fascinating about her story is not just the specific idea that struck her that January morning, but how she managed to realize it. She paved the way for an entirely new field, but it was a path littered with failures and mistaken assumptions. Her experience is not an exception for realizing intersectional ideas. Since quantity of ideas leads to quality of ideas, we should pursue many ideas. This, however, leads to the inescapable paradox that in order to be successful at the Intersection, we must have many failures. The solution to this paradox is to incorporate failures into our overall execution plan. In other words, we have to execute past our failures. Just ask Deborah Prothrow-Stith.[1]

Failures and Success

I MET PROTHROW-STITH in her office at the Harvard School of Public Health two and a half decades after the incident in the ER. Today, she is associate dean of the school and a star among those looking for strategies to prevent youth violence. Like many of the people I have met for this book, she is full of purpose and energy. Her voice is strong and her manner infectious, and I found myself smiling, worrying, and laughing with her while she told me what happened that night so many years ago.

The Insight

"I fell asleep soon after he left," she says, but she felt a dark sense of foreboding. Although what she had done to help the patient was medically correct, it seemed that more violence and injury was about to happen. Yet, there was no recourse to prevent it. There were no protocols, no procedures. In fact, it seemed strange to even worry about it. After all, what business did a physician, much less a medical student, have worrying about violence prevention? Her job was to stitch 'em up and send 'em out. The police took care of the rest.

But what would have happened if, say, the man had arrived after an attempted suicide or a drug overdose? First they would have pumped his stomach and declared him medically stable, and then determined whether or not he was still a danger to himself. If the man at that point had said, "Now don't go to bed, because I'm going home to take some more pills and I'll be right back here," it would have set in motion a range of systematic interventions. Physicians even have the power to force hospitalization if a patient is deemed a risk to himself or herself. The more Prothrow-Stith thought about it, the more she realized that physicians often got involved with prevention of harm by attempting to change patient behavior. She and other physicians pushed people to wear seatbelts, to eat right, to exercise, to avoid risky sexual behavior, and to avoid many other lifestyle hazards. But at that time they didn't do anything to prevent violence.

Yet it is clear that violence *is* a health hazard. A pretty obvious one at that. But it was an issue addressed primarily by the law enforcement field—health care workers had nothing to do with it. Prothrow-Stith never did find out what happened to her patient that night, but the experience opened her eyes. There *was* an intersection between health care and violence prevention. No one had explored this intersection before, so she decided to take it on.

The Execution

Over the next couple of years Prothrow-Stith applied for grants, submitted proposals, and developed action plans aimed at preventing violence by using a health care–oriented approach. She seems like the type of person who carefully crafts a plan and applies it without a hitch. She is both action-oriented and focused. She is also quite structured, using frameworks and diagrams to explain what is going on. At one point she flips open a page from her book, showing me a graph. "See, this is what traditional law enforcement does well and this is what public health does well," she says, placing her finger on the point where they intersect. In short, she doesn't strike you as the type of person who makes a lot of mistakes. "But we did," she says.

Her assumptions seemed obvious enough: Health care workers routinely interact with youths involved in violence. They should therefore be in a great position to help prevent violence-induced injury and also welcome an approach to do so. Prothrow-Stith quickly concluded that the hospital, or even the ER, was a great place to initiate violence-prevention strategies.

Prothrow-Stith's assumptions started unraveling almost immediately. To most of her colleagues, she made no sense at all. Overwhelmingly, they felt that physicians had no role in preventing violence. She was told "again and again that, as violence was not a disease, medicine could not cure it." She laughs out loud while talking about those days. "They would say things like:

'You are medicalizing a sociological problem.'

'There is nothing you can do against violence.'

'This is not your role.'"

It became clear fast that she could not hope to solicit support from many of her colleagues, as she had expected. As an alternative, she sought out other partners. The church agreed to help. So did the police. "When we spoke with police officers that walked the streets, the ones that actually had to deal with the day-to-day effects of violence, we found a tremendously positive response," Prothrow-Stith says. They understood that their work was a response to violence, but not necessarily a prevention of it.

But Prothrow-Stith kept on making mistakes. Some of those initial failures were part of her learning curve. For example, she first called the program the Boston Program for High Risk Youth. It turns out few kids like to think of themselves as "high risk." Other failures were more fundamental. One of her first attempts at a solution was to apply a standard medical services model to the problem of violence. This meant collaborating with public schools and other community agencies to refer youth to violence prevention services in clinical settings. Prothrow-Stith spent a lot of effort trying to encourage "at-risk" youth to establish long-term relationships with physicians and mental health professionals. But the teenagers were extremely reluctant to go for services in health care facilities—particularly if they didn't think they had a problem in the first place. These kids would go to a hospital if they were in pain or had suffered an injury, but not to learn about violence prevention.

Basically, Prothrow-Stith's approach was not working. So she decided to switch her focus. Rather than asking youth to come to clinics, she decided to go to them. To reach students, Prothrow-Stith developed a curriculum consisting of videotaped (staged) acts of violent encounters, often acted out by the students themselves. As part of the exercise, the students picked out the moments they thought were key for preventing the violent encounter that ensued. For instance, an actor from the school might instigate a fight by shouting, "Are you going to let that jerk stomp on your shoes?" The students would stop the tape and immediately come up with less confrontational responses.

"Hey man, a little dirt on your sneaker is not worth fighting about."

"Take it easy, it was an accident."

These responses may seem basic to you and me, but they aren't to many teenagers. Prothrow-Stith found that, for many students, it was inconceivable not to answer every insult with an escalation. Many students didn't know that there were less risky ways to handle a confrontation.

When Prothrow-Stith and her partners introduced the curriculum at their first test school, they successfully measured a decrease in the level of violence at the school. It was good sign: She was on the right path, finally. While she refined the curriculum, Prothrow-Stith and a colleague were able to hire two trainers to expand their reach. During their peak they educated a hundred students every other month with the violence-prevention curriculum. Some of the kids ultimately headed out and educated others, and in a few cases they even started their own youth violence-prevention centers. Many of these students had been in the "high-risk" group themselves and could now involve their friends and peers. As Prothrow-Stith found out, this is the type of grass-roots effort that can have a marked impact on any community that is struggling with violence.

The National Public Health Conference in 1985 was a turning point for Prothrow-Stith. She was able to offer evidence that public health initiatives could help stem violence, and her curriculum quickly gained national attention. Soon she was developing programs for Boston-area hospitals. As a result, children and youth admitted with violent injuries received prevention assessment and follow-up treatment to reduce risks for further injury. The approach consisted of multidisciplinary teams and was modeled after treating patients with asthma and suicidal behavior. Today it's obvious that violence prevention is a public health issue. The Centers for Disease Control (CDC) in Atlanta, for instance, has an entire center devoted to it and physicians routinely involve authorities to prevent domestic or child abuse.

Prothrow-Stith became a driving force in Boston, and ultimately nationally, in making violence prevention a public health issue. At the peak of the violence epidemic in Boston, one or more juveniles were killed every month. In the mid-nineties, however, the city had a stretch of over two years when not a single youth was killed. Although no such success lies exclusively with any one individual, Deborah Prothrow-Stith and the team she built can comfortably take a good deal of credit for it. Her success made her the first woman, second African American, and youngest person ever to be appointed commissioner of public health for the Commonwealth of Massachusetts. But her story also illustrates a troublesome characteristic of the Intersection: Mistakes are inevitable if you want to succeed.

Get Ready for Failure

PERHAPS THE MOST counterintuitive byproduct from the explosion of ideas at the intersection of fields is the simultaneous rise in failures. "We made many mistakes right from the beginning," says Prothrow-Stith. But could she have pulled it off without them? Highly unlikely. The more ideas you execute, the greater the chance of realizing something truly groundbreaking. But not every one of your ideas will work out. Innovative people, then, experience more failures than their less creative counterparts because they pursue more ideas. It is thus very difficult—indeed, this book argues practically impossible—to realize ideas at the Intersection by flawlessly executing well-defined action plans. Yet this is how most of us are trained to think about strategy and implementation. We are, in fact, *conditioned* to approach any new challenge with questions such as: What is our goal and how will we get there?

If you are about to develop a new product, you will draw up a step-by-step launch plan based on your market research, discussions with engineers, and analysis of customer needs. A scientist goes through a similar exercise while putting together a detailed grant proposal. In it he or she describes the basis for their new experiments, how they intend to structure them, the resources needed, and how long it will take. This thoroughness increases chances for funding and for getting results.

This approach works great for directional innovation but poorly for intersectional ideas. The major difference between a directional idea and an intersectional one is that we know where we are going with the former. We therefore have reasonable expectations for how many customers will buy our new product, or readers will read our new book, or what the research results will be in a particular study. Once we have identified those goals and the critical action steps needed to fulfill them, we can set up a detailed plan, gather the resources necessary, and start executing. Someone good at making ideas happen not only is a whiz at figuring out this execution path, but also moves with relentless determination from one step to the next. Failure in such an instance usually means that we may not have met our expectations fully, but we made it part of the way.

Such an approach, however, presupposes that one understands what needs to get done and in what order. Unfortunately, the Intersection is a place where our understanding of what to do and how to do it is opaque, at best. An intersectional idea can go in any number of directions. We don't know which one will work until we start trying them out. Successful execution of intersectional ideas, then, does not come from planning for success, but planning for failure. It is a counterintuitive idea, but a critical one. Since we cannot rely on past experience to devise a perfect execution path, we must rely on learning what works and what doesn't. Failures and mistakes during such a process are inevitable. To sum it up: What Prothrow-Stith experienced is the norm at the Intersection. But as the next chapter will show, there are ways to prepare for it.

ten

How to Succeed in the Face of Failure

PALM PILOTS AND COUNTERPRODUCTIVE CARROTS

M ISTAKES AND FALSE STARTS are part of the process for making ideas happen at the Intersection. If we hope to innovate, we must factor them into the equation. We must continue executing ideas and move past our failures. But how? What ultimately makes someone like Deborah Prothrow-Stith, or anyone else for that matter, successful at the Intersection? In short, she was willing to

- ➤ Try ideas that fail to find those that won't
- ➤ Reserve resources for trial and error
- ➤ Remain motivated

Try Ideas That Fail to Find Those That Won't

F AILURE IS PART of innovation—get used to it. Easier said than done; it's almost impossible to be comfortable with the notion of embracing failure. Failure is demeaning and disappointing. And it's particularly daunting in a competitive organization where it may result in not only loss of confidence, but also a decline in credibility, and possibly even a demotion or loss of job. It is therefore valuable to look at failure from an organizational perspective, since those dynamics are part of what makes us fearful of it.

The smartest managers and best-trained teams understand that failure is part of innovation, and they therefore expect it to happen a certain percentage of the time.[1] Dean Kamen, the prolific inventor of the dialysis machine and the Segway human transporter, was said to be "displeased if he and his engineers weren't frequently failing in preposterous ways, because impressive failures signified impressive aspirations."[2] Clearly, this is not a typical corporate attitude. Even if managers know that failure encourages future innovation, it is not easy to manage for it. It is much easier to manage for success. After all, if someone has done a job well he or she should be rewarded for it—a pat on the shoulder, a bonus, a raise. People expect to be rewarded when they succeed. But how should we handle failure?

To answer this, let's analyze a few work situations. For example, rewarding success alone is a reasonable strategy in jobs where the sole purpose is to execute a specific process: landing an airplane, conducting a surgery, or installing a hard drive. In most cases, these jobs have clearly established processes, and failure is neither expected nor can it be viewed as positive. But what about jobs where success depends on a steady output of new ideas? Those where trial and experimentation are part of the job description? This is trickier. It still seems like a good idea to reward success. But is it enough to reward success and let the failures "slide"? Would we then be willing to risk failure, and therefore increase our odds of innovating in the long term? Maybe. But the best results would come in an environment where success and failure are rewarded equally—and where *inaction* is punished.

What clearly should not be rewarded is doing nothing—not executing any creative ideas at all. Robert Sutton, professor at Stanford Business School, suggests that inaction is far worse than failure in terms of assessing innovative effort. Failure, after all, implies some sort of output. Since the quality of innovation is linked to quantity of ideas, it makes sense to manage according to metrics based on quantity of ideas. Examples of such metrics include the number of prototypes built, patents filed, papers published, projects completed, and so on. Without quantity of ideas, there can be no innovation. Therefore output, whether generating success or failure, must be rewarded.[3]

This may seem unrealistic without additional action items. So how do you reward failure? Sutton has a couple of pointers on how to navigate this terrain.

> Make sure people are aware that failure to execute ideas is the greatest failure, and that it will be punished.
> Make sure everyone learns from past failures; do *not* reward the same mistakes over and over again.
> If people show low failure rates, be suspicious. Maybe they are not taking enough risks, or maybe they are hiding their mistakes, rather than allowing others in the organization to learn from them.
> Hire people who have had intelligent failures and let others in the organization know that's one reason they were hired.

Vertex Pharmaceuticals is a great example of a company that tries ideas that fail in order to find others that won't. Their industry is facing some serious challenges. Although spending on research and development went up more than 300 percent industrywide during the nineties, the number of drugs approved by the Food and Drug Administration (FDA) during that period dropped by 50 percent. Companies are seeing drugs that cost hundreds of millions of dollars to develop stall or fail in the FDA approval process. If this attrition doesn't improve, drug development will become prohibitively expensive for most companies.[4]

Paradoxically, Vertex has attempted to decrease the attrition rate by, in effect, increasing it. The more ideas (molecular combinations in this case) researchers can test, the better their chance of finding a few good ones. Consequently, many more ideas will lead to dead ends, but hopefully at a much earlier stage than before. Good ideas—in this case, effective, safe drugs—have a better chance of getting approved and ultimately generating sales. Vertex has purposefully placed itself at an intersection of disciplines and technologies where it generates thousands of new drug candidates every day.[5] According to Vicki Sato, the president of Vertex, part of the company's strategy is revealed in its name. "You know, vertex does not just mean summit or apex, although that is what most people think of when they hear our name. It also means intersection," she explains.[6]

The goal is to produce as many potential drug compounds as possible in order to select the few that will provide Vertex with a breakthrough drug. This is how it works: Processor-heavy computers, more or less randomly, combine molecules with different drug targets. They then throw out the combinations deemed to be ineffective.[7] The remaining molecule combinations go to a team of computer scientists, biologists, chemists, medical doctors, manufacturers, and lawyers who work together to evaluate them and bring those with the highest potential to fruition. Some compounds are discarded quickly, others a bit later. But some get developed into drugs. On any given day, Vertex's computers generate thousands of combinations; the vast majority of these end up as failures. But by increasing the output of ideas and failing often and quickly, Vertex also has a better chance of developing successful drugs. So far the strategy seems to be working—two drugs have FDA approval and six more are in phase 1 or higher.

This principle holds true not just for companies, but also for individuals. During his work on *The Waste Land*, for instance, T. S. Eliot tested hundreds of unique ideas while writing the poem. Many of them were ultimately discarded; others made the cut. The resulting poem is a masterpiece where Eliot stepped into the intersection of worldwide cultures and mythologies. It took several years to complete and involved numerous rewrites and edits (with significant help from his wife as well as from his friend Ezra Pound). Although the poem may seem like a single product, it was actually a collection of hundreds of different concept combinations.[8]

In order for this strategy to work, it is important to learn from past mistakes. Why is it, for instance, that some individuals or teams come up with a lot of ideas and products, but still cannot achieve relevance or breakthrough? It could be that they have just been unlucky. After all, we know that the likelihood of quality increases with high output, but it's not guaranteed. It is more likely, however, that unsuccessful mass producers are not pursuing different ideas, but are simply producing incremental variations of similar (and mistaken) ideas. Imagine, for instance, writing fifteen very similar books on a topic no one finds worthwhile or valuable. It is all right to make new mistakes, but not to repeat old ones.

In order to successfully realize ideas at the Intersection, you must try out many different ideas generated through the explosion of concept combinations. Some of those ideas will fail . . . the others won't.

Reserve Resources for Many Trials

WHEN THE PALM PILOT was released in April 1996, it became the fastest-selling computer product in history. It was small, sleek, elegant, and irresistibly useful. "I remember sitting around the table when it was presented," said the marketing manager at the time. "We all had goosebumps."[9]

How did the people behind the Palm Pilot, Jeff Hawkins and Donna Dubinsky, achieve such success? Through careful planning and execution? Well, they did plan, but their plans did not work out at first. Before they invented the Palm Pilot, their start-up had launched another handheld device called the Zoomer. The Zoomer was designed to do everything a computer could, even print and fax, despite its compact size. But in the end the Zoomer couldn't do anything a computer could, or at least none of it well. The software barely worked, and the device was too big and too slow. And, of course, no one really wanted to send a fax from a handheld. The Zoomer was a huge failure.

The story might have ended there, but Dubinsky had been careful to reserve enough of their capital to have another go. They realized that people didn't want their handheld devices to emulate their computers. Instead, most people wanted something that was simple enough to compete with their day planners. Something small enough to fit in a shirt pocket and easy to work with—a tool that could accomplish a few important tasks fast. That is ultimately what the Palm Pilot, one of the most innovative products of the nineties, became.

One characteristic of intersectional ideas is that many assumptions you make during development will be wrong. This is why you must not only expect failures but also plan for them. Deborah Prothrow-Stith did it. So did Hawkins and Dubinsky. Anyone succeeding at the Intersection will tell you the same story; their original idea had to be modified again and again. Picasso, for instance, used up no less than *eight notebooks* just for preliminary sketches of his revolutionary painting, *Les Demoiselles d'Avignon*.[10]

This approach, however, requires a careful preservation of resources, whether those resources are money, time, reputation, contacts, or power. Clayton Christensen is a professor at Harvard Business School and a renowned authority on disruptive innovation, a particular type of intersectional idea. He notes in his best-selling book *The Innovator's Dilemma*:

> *Research has shown, in fact, that the vast majority of successful new business ventures abandoned their original business strategies when they began implementing their initial plans and learned what would and would not work in the market. The dominant difference between successful ventures and failed ones,*

generally, is not the astuteness of their original strategy. Guessing the right strategy at the outset is not nearly as important to success as conserving enough resources (or having relationships with trusted backers or investors) so that new business initiatives get a second or third stab at getting it right. Those that run out of resources or credibility before they can iterate towards a new strategy are the ones that will fail.[11]

This seems like smart, straightforward advice. Just keep your purse strings tight and you will be all right. Yet individuals, teams, and companies often run out of resources before they have the chance to fully explore the different roads at the Intersection. Why is that?

When we set up an execution plan we do so for several reasons. We do it to coordinate activities; we do it to plan resources; and we do it to convince partners, investors, customers, or distributors to join us. Most of the time, however, most of these people want to see certainty in our plans. Ending a presentation with the words "but this could all change tomorrow" usually won't win any points. If you are working on a directional idea, it shouldn't. But if you are working on an intersectional idea, you must be wary of substituting inherent uncertainty with concrete plans.

Our desire to predict every detail derives, in part, from a belief that we can eliminate uncertainties with careful planning. Even if we know that the future is uncertain, it may still *feel* like those uncertainties can be controlled if we have worked out the details. But planning at the Intersection is tricky—yes, plans serve a useful purpose, but only if we are aware that they may need to change.

Here is the rub: Unless we tell others that our plans can change, they will form expectations based on them. Instead of being prepared to provide more money if the first trial does not succeed, investors will want to see results. Customers will want things to work according to plan, and our friends and colleagues begin to believe "we are real close to something happening." We, of course, respond to these expectations by trying to get it right the first time. We spend more time, money, and goodwill on executing the wrong plan. And suddenly we run out of resources before we have a chance to adapt and make our intersectional ideas real.

One of the reasons so many Internet start-ups failed so spectacularly can be traced to this issue. During the Internet boom, small start-ups received unprecedented amounts of capital. In many cases they were attempting to do something entirely original, something no one had tried before. We know today that a large number of predictions and assumptions about the Internet were dead wrong. But there is, in fact, nothing unique about that. Being vastly off the mark is nothing special for a start-up at the Intersection. In fact, as Christensen pointed out, it is the *rule* and not the exception.

So if being wrong is not unusual, why did so many Internet companies crash and burn? Because most of them executed *as if they were going to get it right on the first try*. They proceeded as if they didn't think they would have to change their plan once, twice, or more times. Boo.com, for instance, was a London-based fashion company that attempted to merge the global reach of the Internet with sports clothing. It opened for business in November 1999 and crashed less than 7 months later, after burning through at least $135 million. The founders developed a business plan that detailed a worldwide growth strategy, which they sold to investors, suppliers, and customers. They placed their entire bet on that one approach, rather than reserving resources for a couple of trials of different ideas. So Boo launched a global campaign, hired more than 350 people in five offices around the world, and occasionally flew the Concorde to make it to meetings on time.[12]

By the time Boo realized that its initial plan did not work, it was too late for the founders to change course. With money gone and public trust shattered, the collapse was inevitable. But their general idea was not necessarily a bad one. There is clearly a market for online retail clothing. Boo had a good shot at innovating this market and becoming a leader in it. But the founders believed in their business plan enough to place all their resources behind it. Although that might be the way to go for most directional ideas, it is just plain wrong at the Intersection.

How, then, do we combat the desire to spend resources according to plan, knowing that those plans may have to change? What are the lessons we can take away from Deborah Prothrow-Stith and others who have succeeded in realizing intersectional ideas?

- ➤ Be prepared to change your execution plans. You may have drawn them to convince others, motivate yourself, coordinate activities, or for any number of other reasons. But they will be based on at least some faulty assumptions and will therefore need to be adjusted.
- ➤ If realizing your ideas depends on money, make sure you spend it carefully. Is it possible to reserve enough for at least one or two more attempts? Alternatively, find trusted backers who will provide money for several trials.
- ➤ If realizing your idea depends on time, give yourself enough time for several trials and errors.
- ➤ Proceed with extreme caution if your reputation, goodwill, or contacts are riding on a successful execution of your idea on the first try.

Remain Motivated

PERHAPS THE MOST important strategy for success at the Intersection is to remain motivated. If you stay motivated, you will have the wherewithal to push past your mistakes and stick with an idea until you become successful. If you lose this motivation, though, complete failure seems all but inevitable. Not only will you lose interest in what you are doing, but your willingness to explore different creative ideas or to take risks drops quickly. Motivation, then, is crucial for helping you persevere when initial ideas fail.

Great advice, but how do you act on it? One of the most common tools for keeping ourselves and others motivated is incentives. Wouldn't it make sense that this approach should work at the Intersection? After all, rewards have a long and strong history of successfully influencing productive behavior. In fact, one of the most well known psychological experiments, the Skinner Box, illustrates the power of rewards. In it a rat is placed in a box containing a button and a food dispenser. If the rat steps on the button, it will get a reward—in this case, food pellets. The reward, in turn, drives the rat to keep stepping on the button to get more food.

This "Skinner effect" has become the flagship experiment verifying that rewards are a key to behavioral control. If a particular type of behavior is rewarded, then that behavior will be repeated and improved. It is virtually impossible to avoid having personal experience with the Skinner effect. If you want children to mow the lawn and clean up their room, they will do it much more readily if they receive a reward, such as their weekly allowance. Of course, the same holds true for adults. There is nothing particularly curious or strange about this behavior. If the tasks and goals are relatively straightforward, it may be a great idea to make sure that the people engaged are acutely aware of their external rewards, whether they be money or status or fame. But how does this work when the goals are not clear and when we are not sure exactly which steps need to be taken and in which order? In other words, how does it work at the Intersection?

Motivation at the Intersection

Not well, is the short answer. Harvard Business School psychologist Teresa Amabile set out to examine the effects of reward on creativity in a study involving more than a hundred children. The experimenter told the children that she had two different activities for them. One of those activities was to tell a story from a children's book called *A Boy, a Dog, and a Frog*. The book contains thirty colorful pages of pictures with no words. It therefore leaves plenty of room for the children to interpret for themselves what is going on. The other activity involved a Polaroid camera, about which the children were all very curious. Before the experiment started the kids were divided into two separate groups. Children in the first group were told that they could play with the camera now if they promised to tell a story from the book later. Before they could play with the camera, though, they had to sign a contract promising that they would tell the story when they were done. Once the children had taken a couple of pictures they were reminded of their promise, and the experimenter proceeded to the book.[13]

Children in the second group were told that there were two activities for them, but neither was made dependent on the other. Instead, they were simply asked if they wanted to play with the camera (which they did) and then asked to tell a story from the book. There was no promise made nor was any contract signed. Every story from both groups was tape-recorded, transcribed, and later independently rated for creativity by three elementary school teachers. The results were intriguing. They showed clearly that the first group, which played with the camera as a reward for telling a story later, was significantly less creative than the second group. The children did the same activities in the same order, but with vastly different results. Amabile writes in *Creativity in Context*: "The only difference in experiences of rewarded and non-rewarded children in this study was their perception of the reward as contingent or not contingent upon the target activity. Thus it appears that the perception of a task as a means to an end is crucial to creativity decrements in task engagements."

Put another way; just by *saying* that one activity is a reward for another activity can lead to a decrease in actual creative output. When people feel that they are being rewarded for an activity, that feeling of external control is enough to actually impair creativity. Amabile and others have verified the negative effect a reward can have on creativity in numerous studies. Consider yet another experiment: The subjects were asked to mount a candle on a vertical screen. They could use only the screen, the candle, a book of matches, and a box of thumbtacks to solve the problem. This experiment contained what researchers call a "break set," which is a fancy way of saying that the subject must use an object in an unusual way. In this case the subject had to empty the tacks from the box and then tack the box to the screen as a platform for the candle. Of course, the hard part here was seeing that the box could be used as a platform and not merely a container for the tacks. The subjects were then divided into two groups. One group was told that they would receive a $5 reward if their solution time to the problem was in the top quartile and $20 if their solution was the fastest. The second group did not get these instructions. As you might have guessed by now, the group that had no chance of getting a reward solved the problem significantly faster than the people who did.

Explicit rewards, then, can be an effective way to kill off our creativity. Why, exactly? Amabile has found a connection between our internal drive, or intrinsic motivation as she calls it, and our creative efforts. If intrinsic motivation is high, if we are passionate about what we are doing, creativity will flow. External expectations and rewards can kill intrinsic motivation and thus kill creativity. When intrinsic motivation drops off, so does our willingness to explore new avenues and different ideas, something that is crucial at the Intersection. This means that in order to stay motivated and execute an intersectional idea, as did Prothrow-Stith and Hawkins and Dubinsky, we must be careful of explicit, external rewards. Stephen King puts it this way: "Money is great stuff to have, but when it comes to the act of creation, the best thing is not to think of money too much. It constipates the whole process."[14]

Overshadowing Internal Drive

What happens when explicit rewards become so incredibly strong that they overshadow internal motivation? I believe we saw this happen on a mass scale during the Internet boom. I vividly remember taking a tour bus at one of the year's biggest Internet-industry conferences. Next to me sat a guy who wore the New Economy fashion of the time, a stark blue shirt, blazer, and slacks. He had recently graduated from a top-tier business school and taken a job at an investment bank, but left that position to join a start-up. When I asked him what the company did he said it was a B2B, meaning specifically business-to-business interactions over the Internet.

B2B Internet exchanges were all the rage back then. Mind-boggling revenues were predicted, and B2Bs could easily establish billion-dollar valuations, although they were years away from any profit. Around the fall of 2000 there were B2B exchanges for every conceivable industry. So I asked this guy exactly what his B2B firm did. "It is a place for buyers and sellers of fish to meet," he said. They had not yet closed their first round of funding, but were forging ahead with the seed capital they had garnered.

This seemed amazing to me. What could possibly entice someone who lived the fast life on Wall Street to turn his attention toward fish trading, of all things? This is an industry that must have had almost zero appeal to most MBAs at the time. Yet, I knew from the trade magazines that he had at least three or four current competitors and that more were on the way. To trade fish!

The answer is not a big mystery. There were three letters that provided as strong a motivation as any: I-P-O. Going public and becoming instantly wealthy on paper was an intense motivating factor. Amazingly, however, creativity in the B2B segment, which sat at the intersection of splinter technology and established markets, was quite limited. Business magazines even published how-to lists for starting a B2B and taking it public. And everyone followed the formula. Everyone "knew what to do"—it was just a matter of doing it faster than everybody else.

In other words, people were behaving as if they were pursuing directional innovation. Failures were not expected, and when they inevitably happened and the prospects of an IPO grew distant, people had problems staying motivated. This is not to say that the teams embarking upon these adventures could not be creative or entrepreneurial. But it says that intrinsic motivation, a key driver for innovation at the Intersection, can get pummeled when external motivation competes for its attention.

There is other research that indicates just how pervasive this principle is in the corporate world. Jim Collins, author of the best-seller

Good to Great, looked at what type of leaders head up stellar companies and how these leaders are compensated. It turned out that firms used a wide range of incentive schemes: salaries, stock options, bonuses, profit sharing, and so forth, in every single variation, but none of this variation was correlated to success. Incentives were important in attracting a candidate to accept a particular job, but once on the job it hardly mattered at all. People who are driven to perform do so based on internal drive, not on external incentives. They *want* to do a good job.[15]

Still, firms that hope to generate groundbreaking innovation implement all kinds of rewards to make it happen. John Seely Brown, the former director of Xerox's famous Palo Alto Research Center and one of the most respected innovation thought leaders in the country, gave me his take on why corporations seem to ignore the negative effects rewards have on intrinsic motivation. "The reason corporations ignore the truth about passion is because they rely on predictability. Quarter-by-quarter expectations require predictability. In order to be predictable, the management needs to have control. In order to exert control, one uses incentives. One of the most powerful incentives is salaries and bonuses. But innovation doesn't work like that."[16]

It is important to add that not all rewards will have a negative effect on intrinsic motivation. Innovators at the Intersection find that their intrinsic motivation declines when correlated with *explicit* reward, but rewards that are provided as a testament to their competence or as part of a learning experience can prove very effective. This basically means that an innovator should receive the fruits of his or her labor. In fact, if such rewards are not given, this is almost sure to stifle motivation. Amabile notes that "negative effects also seem to appear where people feel they are not being generously and equitably compensated for their work."[17]

Clearly, if someone else gets a reward for work you or your team have done, it will have a negative impact on motivation. Håkan Lans, for instance, who has had to fight many patent battles in court to protect his licensing rights, says, "I do need recognition for what I have done and I do wish to be rewarded for it. If you don't get it, or if this reward is taken away from you, it is demeaning and it hurts. And it can kill the desire to innovate."[18] People at the Intersection must believe that they will get the reward they deserve for their work—even though no one at the outset knows exactly what the reward will look like.

Moving Toward Success

DEBORAH PROTHROW-STITH was able to succeed at the Intersection because she incorporated her failures into the overall execution plan. Most important, she managed to stay motivated throughout the entire trial-and-error process because she did something she truly enjoyed. The same, by the way, holds true for every other person I talked to for this book.

Simply executing a plan past your failures, however, is not enough to succeed with an intersectional idea. There are more challenges to overcome. Not only will you need to face down uncertainty; you will also have to fight the seductive urge to remain comfortably within your established network. In fact, many of the resources, processes, and people that made you successful in the past may suddenly be holding you back. In the following chapter we will look at why.

eleven

Break Out of Your Network

ANTS AND TRUCK DRIVERS

IN THE EARLY 1990s Eric Bonabeau, an R&D engineer at France Telecom, and Guy Theraulaz, an ecologist studying social insects, met at a seminar held by the Santa Fe Institute in New Mexico. They talked about, among other things, how ants find food.[1] Ten years later the techniques that were developed based on this conversation are helping petrol truck drivers plan their routes through the Swiss Alps. Now what, exactly, do foraging ants, Swiss petrol truck drivers, and telecommunications engineers have in common?

For starters, none of them wants to waste time. But while the insects find the quickest path to their destination, human counterparts often take longer routes. How do the ants pull it off? In many ant species, special foragers are sent out along more or less random paths to search for food. Each foraging ant releases a pheromone during its frenetic search, which has the quality of attracting other ants. The stronger the smell, the more ants it attracts. The ant that finds the quickest path from the nest to the food will have the strongest-smelling trail since it returned sooner. The stronger scent, in turn, leads other ants to choose this particular path, and over time it becomes the dominant pheromone trail. In the end, through the collective behavior of the ant colony, the quickest path will have emerged, creating a highway of foragers efficiently ravaging the food source.[2]

When Bonabeau heard Theraulaz give this explanation, he experienced a major *aha*, "not only because I finally understood how ants were able to so efficiently raid my sandwich during these distressing picnics of my childhood, but also because I saw a powerful computing metaphor." The life of an ant colony, it turned out, had connections to other problems in our world, problems Bonabeau had been working on. Suddenly he understood how they related.

"Back at France Telecom," he says, "I started working on applying the ant metaphor to routing, a recurrent telecommunications network problem. Routing is needed because most large-scale communications networks are not fully connected for cost effectiveness, so messages have to be guided through the network to reach their destination. I found that by letting virtual ants leave virtual pheromones at the network's nodes or routers, the routes that messages use can be optimized."[3]

The success of applying insect behavior to computer search algorithms intrigued Bonabeau. "But France Telecom wasn't ready for that," he says with his distinct French accent, "and at the same time I became more and more interested in insects." One can understand the tension. Studying ant pheromones was most likely not very high on the telecom giant's priority list. Bonabeau eventually decided to leave his company and headed straight to the Santa Fe Institute. "About one year later some other people at France Telecom started to wonder whatever had happened to me when they saw a *BusinessWeek* cover story on Peter Cochrane, then head of British Telecom Labs, bragging about using antlike agents for routing in telecom networks." Bonabeau's ideas had spread from the institute and British Telecom had the good sense to pick up on them.[4]

Bonabeau spent three years at the Santa Fe Institute, learning more about wasps and ants than he had ever expected. During this time he pursued this new area with confidence, force, and little consideration for where it would lead him. "I didn't think about the future," he says. "I didn't think about my career or whether what I did would be useful. I didn't think about anything like that." He just stormed ahead, searching for connections between the two fields. He found them.

Eric Bonabeau's insight and efforts ultimately launched an entirely new discipline called "swarm intelligence."[5] It is a fascinating field filled with biologists, computer programmers, and others trying to find trends and answers by running programs that mimic the behavior of social insects. Even Michael Crichton combined the concepts from swarm intelligence with standard thriller fare in his book *Prey*.

Today Bonabeau is chief scientist at Icosystem, a company he founded that applies this type of science to large-scale business problems such as factory scheduling, control systems, and telecommunications routing.[6] He has, for instance, worked with the U.S. Department of Defense to increase the effectiveness of unmanned aerial vehicles (UAVs). This type of airplane became well known to the public during the U.S. war against the Taliban in Afghanistan in 2001. UAVs have no pilots and can, relatively risk free, search vast areas of land for enemy holdouts. The problem is that as the number of UAVs in the air increases, the task of managing this search process quickly becomes inefficient. It is virtually impossible for the unmanned airplanes not to duplicate each other's searches. How can one UAV avoid circling over the same swath of land another UAV just searched? To solve this problem, Bonabeau provided each UAV with a trail of virtual pheromones that told other UAVs to "stay away" for some time. This way the swarm can effectively survey enemy territory.[7]

What about the Swiss truck drivers who deliver fuel? They have the challenge of finding the shortest route between fuel stops in the Alps, and their efficiency varies enormously depending on the order in which they visit them. This problem may seem simple to solve, but with a large number of gasoline deliveries the possible variations in paths quickly becomes impossibly large to analyze. After only a dozen or so you are talking about billions of potential routes, increasing exponentially for each location a driver has to visit. No computer can figure out quickly enough which of those paths is the shortest. Ants, however, have spent millions of evolutionary years doing it. So the drivers use software that mimics the ants' foraging behavior.

Today, Eric Bonabeau has become a pioneer and leader in the field of swarm intelligence. But how did he manage such a thing? How was he able to realize his ideas at this intersection? It happened not only because he executed past his failures (he's had his share of those), but also because he dared to break away from his carefully constructed environment. It happened because Bonabeau was prepared to break out of his old network in order to build a new one.

The Network Paradox

W HAT HAPPENS when you decide to pursue an intersectional idea? Say that you have managed to break down the associative boundaries between two (or more) different fields and you have also managed to ignite an explosion of randomly combined concepts. Suppose that the only thing standing between you and success at the Intersection is to apply your ideas. So you proceed, prepared to execute your vision undeterred by early failures. But suddenly you discover something unexpected. Virtually all of your existing relationships and structures seem to be holding you back. Colleagues, career track, mentors, institutions, customers, traditions, peers, distributors, suppliers— all those people and things that helped you succeed in the past seem to be conspiring to keep you in neutral. They urge you to stay within your own field—away from the Intersection.

It is not that the network is holding you back on purpose. There is no conspiracy. But your network will promote, support, and highlight ideas that are valued within it. And it squashes or removes ideas that are not. This inherent characteristic creates a difficult paradox for anyone pursuing an intersectional idea: If we wish to succeed at the intersection of fields, we have to break away from the very networks that made us successful. Why is it that some of the greatest obstacles on the road to the Intersection can come from people within our own disciplines, from our customers, our organizations, our own cultures, or our other close relationships? To answer these questions, we must understand why we build networks in the first place.

The Reason We Build Networks

W HY DO WE GET BETTER at generating and executing ideas within a field? Although it is clear that an increase in experience (leading to an increase in understanding of concepts) is an important factor, many other factors contribute to a person's or a firm's success within a field. We become successful because we have strong relationships with business partners and mentors, because we understand what our customers and employees want from us, because we share many of the same goals as our company or institution, and because we have learned from departments and colleagues what it takes to succeed. These relationships become deeply integrated into a tight network supporting similar values.

Clayton Christensen calls these types of networks *value networks*. By this he means "the context in which a firm identifies and responds to customers' needs, solves problems, procures input, reacts to competitors and strives for profit."[8] Values, by this definition, have nothing to do with integrity or morals. What Christensen talks about is much more pragmatic. Within this context two firms share the same values if they, say, value sales over profit, design over functionality, size over speed, and so on.

Christensen studied the value network for a number of industries and noticed that firms have to develop value networks in order to succeed. Consider the disk-drive market, as Christensen did in *The Innovator's Dilemma*. Disk drives are always included as a component in another product, such as a desktop computer or a laptop. This means that a firm that produces disk drives must be very much in synch with the firms that make the computers. Any changes made in this system require all parties to move more or less in tandem. This inevitably leads the firms to develop a common understanding of what is important, of what is valued. Those firms operating in a laptop value network, for instance, will value efficient energy use and small size more than those operating within a desktop value network. These values will color everything that happens in these organizations, from the promotion of new ideas to the allocation of resources. They will therefore greatly affect the people in it.

"As firms gain experience within a given network, they are likely to develop capabilities, organizational structures, and cultures tailored to their value network's distinctive requirements," Christensen writes. Firms, for instance, develop business practices that conform to a particular customer need. If Harley-Davidson bikers want their motorbikes to be big and powerful, then Harley distributors want the same, as do the suppliers, as does Harley-Davidson itself. Harley will therefore hire people who like their bikes that way, they will develop a culture that prefers bikes that way, and the culture will promote ideas that are based on bikes remaining that way.

Although Christensen was focused mainly on corporations, it is easy to see how his argument also applies to individuals. People become part of larger networks built on interlocking relationships, just like disk-drive manufacturers. In order to succeed within a field people will acquire specific experience, strike alliances and partnerships with others, and align themselves with firms, organizations, or institutions that support the values within the field. This clearly holds true for someone like Eric Bonabeau, who worked as an R&D engineer for France Telecom, a large and bureaucratic organization. But it holds true even for those people we believe are much freer to pursue different types of projects, such as researchers at universities, serial entrepreneurs, and artists.

Imagine what the value network of a decently successful musician could look like. She will have acquired hard-earned skills with certain instruments. She will have developed great relationships with her band members, producers, and distributors—all of them creating and selling a particular brand of music. Together they will have established contacts with editors and executives in music media, and they will have relationships with club and show owners. These are people who understand her music and can promote it. Most important, the artist will have an established fan base, people who can be counted on to buy new CDs or attend concerts. All of these individuals, firms, and customers form a value network around a particular type of music—and the artist is caught in the middle. Her situation is not necessarily all that different from that of a major corporation.

Value networks are needed to succeed within a field. That's why we form them. And that is, as you may have guessed, where all the trouble starts.

Why We Have to Break Away from Networks

ALTHOUGH VALUE NETWORKS are essential for directional innovation, they can prevent us from successfully pursuing intersectional innovation. Christensen pinpointed value networks as the main culprit for why great firms, those "that have their competitive antennae up, listen astutely to their customers, [and] invest aggressively in new technologies," ultimately fail.[9] In order to succeed within their field (which they may even have created!), they had built strong value networks. Unfortunately, those value networks actively work against the firm's desire to realize intersectional ideas. This ultimately leaves such firms open to attack from upstarts that can innovate past them.

Remember the discussion of the animation industry back in chapter 2? With movies such as *Finding Nemo* and *Monsters, Inc.*, companies like Pixar have used computer-generated graphics to upend the traditional 2D animation market. Looking back on this example, one can see that this was a classic case of the traditional animation studios getting caught in their value network.

The making of a full-length traditional 2D animation feature, after all, requires a lot of skill and manpower, and the Walt Disney Company has been the notable leader in this field ever since the days of *Snow White*. Traditional animation has been its heart and soul—the people it hires have been trained in this field, major investment decisions have been made with this in mind, the Disney organization is built around it, and even during the 1990s this image was supported by blockbuster traditional animation movies such as *Beauty and the Beast*, *Aladdin*, and *The Lion King*.

At one point, however, Disney became interested in delving into computer graphics, and it forged ahead with the 1982 feature *Tron*. Although *Tron* broke new ground in the area of computer-generated effects, it did not perform very well in the box office. But Disney's value network demanded large-scale successful feature movies. To generate those, they required better technologies for more interesting special effects and a large staff who could produce them. As a result of their experience with *Tron* and the limited resources dedicated to the genre, they concluded that their foray into computer graphics should be halted. Logical? Yes, but only because the studio decided to remain aligned with its established value network.

A firm unencumbered by established value networks, on the other hand, might look at the situation differently. Pixar was formed when Steve Jobs bought the small computer graphics unit from Lucasfilm and applied its technology to the field of animation. Although the team was not ready yet for feature-length films, it found smaller markets that would benefit from its intersectional approach. Between 1986, when Pixar was formed, and 1994, Pixar produced nothing but short films and commercials.[10] Disney's value network, in contrast, would have prohibited them from pursuing those types of projects.

Once Pixar stepped into the Intersection, however, it could directionally improve its innovation and slowly move into other markets. Over time the technology and skill level became good enough for use in feature-length animation movies—and when it produced one, it had a blast-out success. Interestingly, Disney decided to partner with Pixar in order to keep a footing in the new techniques. This was, of course, a lot easier than breaking away from its old value network. But when Pixar ultimately decided to leave the partnership in 2004, Disney had to do it anyway.

A business inside a value network will have a much tougher time pursuing an intersectional idea than one that has managed to remain outside of it. So will a person. Let's return to our musician for a moment. Let's say that this artist has broken away from her usual sound and composed music that combines elements from several different fields. It is unlike anything else, and it could conceivably launch an entirely new genre of artists and sounds. What would happen if she attempted to pursue this idea?

First, she would probably have to develop some new artistic skills. Next, she might have to leave her old band members behind and search for other musicians with the skills to execute this new type of music. She would also soon realize that all of her relationships with distributors, producers, media editors, and executives are entrenched in her former genre. These relationships took years to develop. When approached, these people might not respond to this new type of music. So who will? Well, she doesn't know since this type of music doesn't yet exist. Finally, her fan base, those who buy her albums and go to her concerts, might reject her new CD, snickering that she has "lost it" and hoping that she quickly returns to the music she knows and they like. Sales of the new album could be horrible, at least compared to sales of her previous recordings. With such a lack of predictable success and continuing uncertainty about the future, she may simply abandon this new sound and return to what she knows will work. We all face similar challenges when executing an intersectional idea.

Both people and firms in a value network will have set up processes and procedures that essentially kill off attempts to break out of it. New ideas that do not correspond to the values of the network have a way of getting eliminated. This is why we must break out of these networks if we want to enter the Intersection with the highest chance of success. That is what Bonabeau did. He left France Telecom and headed to the Santa Fe Institute. He formed relationships with a new set of researchers, institutions, peers, and customers for his ideas. He succeeded at the Intersection because he broke out of his old network and built a new one. This is not easy, but as the following chapter will show, it is possible.

CHAPTER twelve

How to Leave the Network Behind

PENGUINS AND MEDITATION

I s t h e r e a n y w a y we can avoid the need to break out of established value networks when we step into the Intersection? Yes, but only if we have not built them in the first place. For example, at the start of a person's career, there has been little chance to form any. In addition, if you have led an incredibly diverse career, routinely going from one field to another, you may not have had enough time to develop deep networks.

This, however, is a very rare scenario. Most of us have built networks around an established field. How do we escape from the networks that once were so helpful to us? The following two strategies can help:

➤ Break the chain of dependence
➤ Prepare for a fight

Break the Chain of Dependence

T HE ONLY WAY TO SUCCEED in breaking away from your old value network is to stop relying on it. Sometimes this means that you have to quit your job and join an institution that can quickly help you establish a new network, like Eric Bonabeau did. Other times it means you have to start building new relationships almost from scratch. Building a new network means being prepared to find new colleagues, new organizations, and new buyers of your ideas and products. Your immediate reaction may be that if you have to build from scratch, there will be no way to catch up with competitors that have several years of advantage. But remember, you are not catching up with others. At the intersection of fields there are no others, or at least not very many of them.

Some people may believe that breaking away from one's past network is the same as leaving it in the dust. But if you wish to maximize your chances at the Intersection, the last thing you want to do is alienate your allies. Not only is your previous network likely to contain great personal relationships, but it can, and probably will, prove very useful to you in the future. In fact, some aspects of your old network will very likely become part of your new one.

Many people in this book have managed to break away from their established value network without alienating it. Deborah Prothrow-Stith, if you recall, had to leave her physician value network in order to connect violence prevention with health care. Her first success came when she developed school curricula, an area seemingly separate from hospitals. But later she was able to integrate her experiences back into work done at hospitals. She had to break away from her field to persist past her failures, but she later was able to reconnect. Another person who was previously part of a physician's value network is Deepak Chopra. Today he is known as the pioneer who brought together traditional Western medicine and alternative Eastern medicine.

Chopra stepped into this intersection, and for the past two decades has produced a plethora of health-care-related ideas, combining concepts from the two fields. As a result, *Time* magazine named Deepak Chopra one of the top one hundred heroes and icons of the twentieth century, and former Soviet president Mikhail Gorbachev called him "one of the most lucid and inspired philosophers of our time."[1]

I caught up with him just as he was preparing for a month-long trip to India.[2] Chopra has a philosopher's way with words. He says things like, "Nature in its unbroken wholeness is inherently creative." As we spoke it became very clear that he has thought profoundly about creativity and innovation. Deepak Chopra used to be firmly entrenched in a very traditional field of medicine: endocrinology. During the 1980s he worked as the chief of staff at New England Memorial Hospital and later he built an endocrinology practice. Back then Chopra chugged coffee in the morning, smoked cigarettes, and drank whiskey in the evening to relax.

At some point during the eighties he began to suspect that he was missing a major piece of understanding about human health. It happened, he says, "when I started to notice things that couldn't be explained by theory. You start to recognize that the theory is full of loopholes." To Chopra it was obvious that medical interventions could not explain the variations in how different people recover. "Two patients with the same illness receive the same treatment but have different outcomes," he said, or "one hundred people with the same pathogens, but only some get sick; others don't," he continued. He felt there were other things at play to explain these differences. If for instance, we take care of our mental health in addition to our physical well-being, we will be healthier overall. A combination of eating right and meditating, for example, could have positive effects on health. Chopra felt sure that if this "alternative" approach to health care was combined with the technological advances embraced by Western medicine, we could make great strides. He also suspected that the traditional field of medicine was covering up these observations. Ultimately, he concluded that he had more to learn. "I realized the education I had received was incomplete. That the traditional point of view doesn't explain everything."

Chopra started to publish his observations, but his conclusions did not sit well with the established medical community. "I was teaching during this time, and noticed that my peers were embarrassed over me so I quit." It wasn't an easy decision. "I was not sure where my income would come from, but I felt a great excitement for the unknown. Many times it was two steps forward and one back. When I started out people thought I was on some fringe. They thought I was certifiably insane." But Chopra was willing to risk his reputation. "It's the prime principle of creativity: You must take risks. All creativity lies in the unknown, not in the known."

Today, thirty-six books later, he is acknowledged as one of the greatest leaders in the field of mind-body medicine. In 1995 he founded the Chopra Center for Well Being in La Jolla, California, and from there he has had an impact across the entire world.[3] Although his career has had its share of controversy, there is no doubt that his insights have paved the way for a new field, one centered at the intersection of traditional Western and alternative Eastern medicine.[4]

Although Deepak Chopra stopped relying on his past network, he did not alienate it. The very same types of institutions that rejected him in the past are welcoming him now. "Once a year I give the keynote at a Harvard Medical School conference. Just gave one in December," he says and laughs a little. He admits to a certain sense of satisfaction with what he has accomplished in the face of the original misgivings. He adds, "You know, it took fifteen years, but courses that [the Chopra Center for Well Being] offers now get AMA [American Medical Association] credit. Today, even NIH supports research in this field."

The people we've met in this book have, like Chopra, managed to break away from their value networks. Marcus Samuelsson, for instance, let most of the existing chefs at Aquavit go when he needed to start from scratch in exploring his new cuisine; Richard Garfield left his job and career as a math teacher to realize his ideas around Magic; and Håkan Lans never stays for too long with any one company, idea, or industry. The same holds true for corporations. Established companies may have to create independent divisions or even spin off entities to break away from their networks. Take a good look around you and try to spot those things that have become critical pieces of your value network over the years. I am not suggesting that you abandon them, but if you wish to enter the Intersection, you must stop depending on them.

Prepare to Fight

WHEN YOU STEP into the intersection of domains, disciplines, or cultures, you must be prepared to fight a battle. The fight can happen on many levels. People may not believe in what you are doing and will make their doubts crystal clear. Sometimes an intersectional idea can threaten an established field. Those within that field will naturally do whatever they can to prevent your idea from becoming an accepted innovation. In order to break out of your network and start building another one, you will need to stand up to the challenges posed by those within established fields. Håkan Lans fought many times against large corporations and the various stakeholders in fields he broke away from. Deborah Prothrow-Stith and Deepak Chopra had to do the same. So did another innovator: a sandy-haired, self-taught computer hacker named Linus Torvalds.

Torvalds had no intention of taking on the establishment at the age of twenty-one. But he did, or at least his ideas did. And the funny thing is—it sort of just happened. Torvalds created an operating system called Linux, which today is among the fastest growing in the world. Linux is the combination of two concepts: that of a scalable operating system and an "open source" copyright model. Open source in the early nineties was seen as a hacker's equivalent of "free love." It essentially meant that anyone was allowed to use Linux as long as they did not sell it, and anyone was allowed to improve Linux as long as they told the world what improvements they had made. Such a copyright encouraged other hackers to tweak and improve Linux. At first this group of software developers numbered only in the dozens. But it grew. More than Torvalds could have ever imagined. Today "the penguin," which is Linux's logo, has a following in the hundreds of thousands.[5]

Imagine that many people testing and improving a product. Sooner or later the product would have to become very good. That's exactly what happened. When Torvalds released versions of Linux in the early years it was unstable, did not work for most computers, and could even kill someone's hard drive. But no longer: Today many *Fortune* 500 companies use Linux because it is faster and cheaper than the alternatives.[6]

When Torvalds released the first version of Linux back in 1991, it stirred quite a bit of excitement in the development community. One person in that world, however, was none too pleased. That person was Professor Andrew Tanenbaum in the Netherlands. Tanenbaum was the master of operating systems. He had developed minix, an operating system that Linux suddenly threatened with extinction. It didn't take long for Tanenbaum to publicly oppose Linux.

> *As a result of my occupation, I think I know a bit about where operating systems are going in the next decade or so. . . . This is a giant step back into the 1970's. . . . Be thankful you are not my student. You would not get a high grade for such a design :-)*

But Torvalds refused to cede and rebutted the specific criticism. He added some lines that have become part of Internet legend from the early Linux days:

> *Your job is being a professor and researcher: That's one hell of a good excuse for some of the brain-damages of minix.*

This type of public discussion went for a couple more rounds, since neither of them would let up. At one point Tanenbaum said, "Linux is obsolete."[7]

This was the lowest point in the entire project for Torvalds. Professor Tanenbaum had publicly lashed out at his "incompetence." But Torvalds fought back. In the end, he persevered. The battle offered valuable practice, because during the next decade a large number of well-financed firms and individuals challenged Linux. Torvalds fought against early naysayers and, later, against corporations such as Microsoft. Although your confrontations may never be this extreme, you must be prepared to fight those that doubt or fear your explorations at the Intersection. Otherwise you might as well resign yourself to your established field.

I finally realized how far Linux had come when I spotted an ad from Oracle, one of Microsoft's fiercest competitors, on the back of an issue of *The Economist*. "Unbreakable Linux," the ad said, followed by, "Everyone knows Linux costs less. Now it's faster and more reliable too." That's amazing, I thought. From hacker utopia to blue chip . . . in less than a decade.

The Missing Piece

ALL OF THE ADVICE offered so far—break out of your field, stop depending on your networks, prepare for a fight—requires something that most of us are uncomfortable with: the ability to live with risk. And risk taking is essential if you wish to turn your intersectional idea into innovation. How do you find the courage to leave an established network behind? To persevere after failures? This is the missing piece for executing ideas at the Intersection, and we will explore it in the following chapter.

thirteen

Take Risks and Overcome Fear

AIRPLANES AND SERIAL
ENTREPRENEURS

B Y 1 9 8 4 Richard Branson had managed to build his record
label Virgin Music into a leading record company with
over 11 million British pounds in profits.[1] In February of that year an
American lawyer named Randolph Fields asked him if he would be
interested in teaming up to start a transatlantic airline. Most people would
probably have scratched their heads, wondering why anyone thought a
music company executive would do such a thing. Branson, however,
was intrigued by the idea and he immediately conducted the following
market research.

First he called the reservation office for People Express, an airline
that offered cheap fares between London and New York. He got a busy
signal and couldn't get through to a customer representative despite
calling all weekend. Branson concluded that People Express either had
a disasterous management team, in which case they could easily be
outcompeted, or that they had more customers than they could handle,
in which case there was room for another competitor. The following

day Branson called Boeing to see if he could lease a jumbo jet for a year, assuming he could return it if this airline thing didn't work out. After a day of shuffling Branson back and forth between different managers, Boeing finally agreed. Armed with this "detailed" analysis, Branson called his business partners.

"What do you think about starting an airline? I've got a proposal here. . . ."

"For God's sake! You're crazy. Come off it."

"I'm serious."

"You're not. You're mad."

Six months later Virgin Atlantic made its first flight from London to New York City. Branson went on to make his airline into a huge success, and it now has flights to cities all around the world. Tens of thousands of people knew more about airplanes, airlines, and travel than Branson. What gave him the confidence to take them on? His market research consisted of only two phone calls, and one of them didn't even go through! Why is it that someone like Richard Branson finds the courage to do what others, who would have a far better chance of success, don't? Branson happened to see a connection between how Virgin ran its music business and how it would run an airline (great customer service). No other entertainment executives, eager to diversify, saw such a connection; but even if they had, would they have dared to pursue it?

There is a common notion that Branson's daring ventures have something to do with who he is as a person. They are part of how he is wired. Taking huge risks is like breathing air for Branson; it's a behavior that's possibly in his genes. This notion suggests that there is little we can learn from Branson's approach to risk taking. All we can do is look at it, shrug our shoulders, and shake our heads at his craziness and then go back to our business. Not so fast.

Branson's antics aren't as crazy as they seem. By studying people like him, we can unearth some important clues about how to face our fears at the intersection of fields.

Risks and Fears at the Intersection

P URSUING ANY VENTURE is filled with some degree of risk. Is there any reason to believe that the stakes in pursuing intersectional ideas are any different from those in pursing directional ideas? There is, and the differences can be significant.

First, the Intersection is unknown territory where we cannot easily apply past knowledge and experience. Within an established field we can estimate our chances of developing a new market, writing another techno-thriller, or pursuing another gene sequence. Failure, as I mentioned in chapter 9, means that although we did not meet expectations, we at least made it part of the way. For instance, sales may not have been as good as we expected, but at least we had some; or it may have taken more time to develop a particular technology than we initially guessed, but at least we're getting there. Either way, we will usually have made progress in the desired direction.

At the Intersection, however, failure can mean that an idea does not work at all. Failure here can be complete. Would it really be possible to use ant colony behavior to make telecom messaging more efficient—or would it just be a gigantic waste of time? And how about Richard Garfield's trading card game—would anyone buy it? No one knew since no one had played anything like it before. The uncertainty is enough to scare off many prospective intersection hunters.

In addition, because of something called acceptable failure, society's expectations can make the perceived stakes at the Intersection seem much higher than those associated with directional ideas. The risk people tend to fear most is not financial loss or wasted time. Rather, it is the risk to their pride, status, and prestige, to what their peers will think of them if they fail.[2] In other words, the risk of failure can weigh more heavily than what is at risk. In this regard, the Intersection, which is an unknown, is at a supreme disadvantage compared to acting in a field that is already generally accepted. If you take a course of action that is widely seen as correct, your reputation barely suffers if you did not make it all the way. If, on the other hand, you proceed in a way that is less understood and fail, it might be tougher to live down because you will be judged harshly. The stigma of failure can be crushing. The economist and "worldly philosopher" John Maynard Keynes put it succinctly: "[I]t is better to fail conventionally than succeed unconventionally."

Although the evidence is mostly anecdotal, there seems to be a clear link between a specific society's stigma of failure and the corresponding amount of entrepreneurial activity. To fail in business in Europe or Asia has more dire repercussions than failing in the United States. In Europe or Asia it could lead not just to severe financial consequences, but also to being "considered a 'loser' by your peers," according to one report by the European Commission.[3] In the United States, however, entrepreneurial failure is often seen as acceptable. Still, even in the United States, if the stigma of failure is a possibility, it can cause individuals to cling to wrong ideas or poor performing ventures longer than necessary, simply to avoid being labeled a loser. Unfortunately, such behavior decreases the likelihood of generating and pursuing other groundbreaking ideas. Fear of failure at the Intersection, then, can be strong. How do we overcome it?

One commonsense strategy I sometimes hear is that we should minimize the risks involved. One way to minimize risks, we imagine, would be to gather more resources than necessary to execute a project. If we had more money, more time, or more contacts, more of everything, we could minimize the risk of failure. But is this notion justified? Is that how risks, fears, and courage work?

How Intersectional Courage Works

H OWARD BERKE is the cofounder and chairman of Konarka, a photovoltaic company that uses new chemical techniques to convert light into electricity through solar power cells.[4] Konarka is, in short, a company that sits at the intersection of energy and chemistry. Something that shouldn't surprise us at this point is that none of the people on the founding team have any real experience in the energy business. Berke certainly doesn't. And cofounder Allen Heeger, who received the 2000 Nobel Prize in chemistry, doesn't either. But isn't it a little surprising that Berke doesn't seem bothered by that fact that solar power companies have an incredibly long history of failure?[5]

Not only does Konarka run the risk of not being able to sell its products—that is a gamble all companies make—but Konarka also faces "huge technology risk as well," according to Berke. In other words, it takes the chance that its technology will not work. "But that is no reason to stop. Many people become prisoners of the fear of failure," he points out, and then adds matter of factly, "If the company fails . . . well, then I'll just have to start another one."

Howard Berke, a soft-spoken yet determined man, actually started another company prior to Konarka. In fact, since the early eighties Berke has founded or cofounded *twelve* companies and has been involved with many more. Serial entrepreneur does not even begin to describe his background. "Let's see," he says, when I ask what happened to them all. "Three of them went to IPO so I guess you could call them quite successful. Three of them were moderately successful, another three were minimally successful. One was a clear disappointment and two of them, well, it's really too early to tell." Berke shows obvious passion for whatever project he is talking about. He likes it best when a company is small enough for its employees to "share a large pizza." His scorecard is impressive, and it may result in part from his pursuit of ideas at the Intersection and his ability to avoid the traps most people face there.

"Each company I started has been at the intersection of at least two industries. It is a deliberate strategy; it is how you innovate." The trail started with ADAC Laboratories more than twenty years ago, a company he did not found but joined early on. "We provided digital imaging for medical diagnostics," Berke says. Today that is old news, but at the time few had made the connection. Ever since then he has taken to "hanging around geniuses" from different disciplines. Although they are not good at understanding how their ideas intersect with other fields, Berke is. "I keep asking them questions. How about this? How about that? Sooner or later we find something that works." His companies have spanned a wide range of areas: medical devices, health care, biotechnology, software, energy technology, communications, and cookies— the ones you bake. "I don't know anything about cookies," Berke says. "So I'm doing a lot of things I have never done before. It's exciting."

He is acutely aware of the risks involved with operating at the intersection of two fields. "If you identify confluence between two industries it can form the basis for a new industry," he says, "but there are risks with that. The risk is that, yes, you can be right, but you can be early. The danger is that the particular fields intersect, but not for another ten years. You get to this place you call the Intersection, but there is no one else at the party." Berke doesn't mind these risks. Actually, he seems to rather enjoy them. He sees little point in creating a company doing something that established businesses already do. Instead, he's driven by the opportunity to innovate. "This way, you at least have a *shot* at being a breakthrough company," he says.

Although Howard Berke is a perfect example of how intersectional courage works, one thing quickly becomes very clear. He may be doing a lot of different things, but none of them are about minimizing risks. Although one might assume that more resources will increase the chances for success, he believes it sometimes lowers them. No, he does not seem to pursue intersectional ideas, or anything else for that matter, by minimizing risks. You may be surprised to learn that you probably don't, either.

People Don't Minimize Risks

HUMANS HAVE a fundamental tendency to live their lives at a certain "acceptable" level of risk. This level is different for each individual, and it changes according to phases of our lives, but we all have a level at which we're comfortable. Gerald Wilde, a Canadian psychologist and leading risk expert, calls this tendency *risk homeostasis*. In a nutshell, risk homeostasis says that people will compensate for taking higher risks in one area of life by taking lower risks in another.[6]

Imagine, for instance, that you are driving your car and you enter a dangerous section of the road, filled with narrow curves and poor lighting. You would naturally slow down to compensate for the new dangers. Conversely, when you leave this risky section and encounter wider, straighter lanes and better lighting, you speed up again. This behavioral pattern makes so much sense that it cannot really be questioned. But the implication of such behavior is anything but intuitive. It suggests that efforts we take to decrease risks around us, such as making roads safer, amount to little because our behavior becomes riskier to compensate.

This was the conclusion in a famous study in Munich, Germany, where researchers installed antilock brake systems (ABS) in half the cars of a taxi fleet, but did nothing to the other half. The drivers were then secretly monitored for three years with hidden sensors. The beauty of ABS is that it gives the driver far better control over the car. It prevents the wheels from locking up under extreme braking conditions and it makes it easier to steer while braking. You would therefore expect that such a brake system would lead to fewer accidents. But it didn't. Drivers with ABS had the same accident rate as those without it, mostly because they drove more aggressively, braked harder, accelerated faster, swerved over lanes, and took sharper corners.[7]

This seems just a little bit counterintuitive. Risk homeostasis suggests that many of the efforts we make to save lives actually don't. Gerald Wilde studied traffic accidents because statistics are well kept, and he found the same patterns wherever he looked. Students who were enrolled in the most elaborate state-of-the-art driver's education programs still had the same number of traffic accidents as those who had only minimal training, because people who are poorer drivers take fewer risks on the road. Crosswalks that were marked with zebra stripes did not decrease the accident rate at the crosswalk, probably because the crosswalks give pedestrians "a false sense of security that the motorist can, and will, stop in all cases," as several reports on the matter concluded.

We all know that wearing a seatbelt will make it much more likely for you to survive a crash. But it will ultimately save lives only if we don't also change our driving behavior while wearing them. Say that you drove a car without seatbelts—would you drive more carefully? Most people would. This, of course, is just the same as saying that we drive less carefully with seatbelts on—a fact borne out by research. Risk homeostasis affects us in ways we can't foresee. For instance, when childproof lids on medicine bottles were introduced, it led to a significant increase in the number of child poisonings because parents became less careful about keeping the bottles away from their children. How does it affect us at the Intersection?

Get Going with Enough Resources, But Not More

FOR ONE THING, risk homeostasis explains why the notion of minimizing risks at the Intersection is not a viable strategy. More money, more time, more experience, or better contacts are all variables that would help in realizing an intersectional idea, we imagine. They do help, obviously, but not necessarily by decreasing the risk of failure. These resources are all factors that can help us in *what* we can accomplish, but they do not increase the *chances* of it succeeding, since with more resources we will try to accomplish more.

In other words, more money leads to greater spending. Having more time means taking more time. Having greater experience or better contacts means relying more on them to get things done. It is not that we waste time, money, or contacts, but that we try to do more with the amount that we have. In trying to do more, we slowly begin to increase the risk of failure, until we hit a level we are subconsciously comfortable with.

What this tells us is that, from a failure perspective, it will not ultimately matter much when you decide to step into the Intersection. Once you have achieved a threshold level of resources, what Berke calls "the minimum amount needed to get your idea going," you should start exploring the Intersection. No point in waiting. Getting more will not decrease the risk of failure. Branson, for instance, figured that he needed an airplane from Boeing and one year's time to see if the airline idea could work. He lacked any experience in running an airline, which in theory raised the risks, but he could work harder to compensate for that fact, which would decrease them again. Get the resources you need (enough for a couple of trials, as discussed in chapter 10), but don't get more.

Berke's and Branson's Secret

I F MINIMIZING RISKS is not the secret behind Berke's and Branson's ability to face fear, what is? Do they just ignore their fears and storm ahead? That doesn't seem to be it either. If you listen closely to what Berke and Branson say, it becomes clear that they do not exactly disregard the fear of failure. Rather, it seems, they accept that failure is part of innovation, and therefore they can somehow embrace it. So what enables them to accept these risks?

They are special because they know a secret, one that I have been giving away throughout this book. The secret is this: If you want to create something revolutionary, head toward the Intersection. The Intersection represents the best chance to innovate because of the explosion of unique concept combinations. It offers a great numerical advantage when looking for fresh ideas. In other words, the Intersection is a low-risk proposition for breaking new ground.

People like Branson and Berke know this. They have therefore avoided the human traps that commonly surround risk taking because these traps dissuade us from pursuing intersectional ideas and encourage us to stay within our field. Branson and Berke have adopted a more balanced view of risk at the Intersection. As you will see, you can do the same.

CHAPTER fourteen

How to Adopt a Balanced View of Risk

ELEPHANTS AND EPIDEMICS

W E HUMANS are not exactly rational in how we think about risk. Emotions, and fear in particular, play a big part in our perception of possible loss versus potential gain. We need to understand more about this aspect of human psychology if we are to understand how people like Richard Branson and Howard Berke behave when facing risks at the Intersection. What are they doing to find the courage others seemingly cannot? And how can we emulate it? There are at least two different strategies we can follow:

> Avoid behavioral traps relating to risk
> Acknowledge risks and fears

Avoid Behavioral Traps Relating to Risk

"Y OU DON'T WANT TO CREATE an environment that becomes overly risk averse," Carly Fiorina, CEO of Hewlett-Packard, told a reporter for an interview in *Fortune*. "Because business is about taking risks. It's about taking prudent risks, calculated risks, but business doesn't happen unless people take some risks."[1]

What does a calculated approach to risk taking look like? This should be a pretty straightforward question to answer. If you were offered an 80 percent chance of making $10,000 or a 50 percent of chance of making the same amount, what would you choose? The answer is obvious—you would go for the gamble with the higher probability of making money. Basically, a calculated approach is about taking the gamble where the chances of winning (and winning more) are the highest. Simple, right?

Yes, in theory. But real life doesn't work like that. First, it is often difficult to calculate the chances of success or failure accurately. At the Intersection it is impossible. Second, winning and losing are not that easy to define when they involve more than just money—factors such as status, happiness, and reputation. Finally, even if we take those things into account, our emotions wreak havoc on how we make choices in the face of risk and fear. That is, even if we know the odds and the outcomes *precisely*, we tend to make irrational decisions, often based on emotions connected with fear. These emotions can affect everyone, even when we are aware it is happening. Peter Bernstein, author of *Against the Gods: The Remarkable Story of Risk*, offers a brilliant example of this behavior in his book.

> One winter night during one of the many German air raids on Moscow in World War II, a distinguished Soviet professor of statistics showed up at his local air-raid shelter. He had never appeared there before. "There are seven million people in Moscow," he used to say. "Why should I expect them to hit me?"

His friends were astonished to see him and asked what had hap-
pened to change his mind. "Look," he explained, "there are seven
million people in Moscow and one elephant. Last night they got
the elephant."[2]

The statistician in this story knew that the chances of getting hit
were remote. Yet because something seemingly improbable happened
the night before, his emotions weighed more heavily than his cold, rational calculation of the numbers. A purely calculated approach, then,
would seem difficult to achieve when it comes to weighing and react-
ing to risk. Human emotional quirks trip us up. These same emotions
make innovating at the intersections of disciplines and cultures more
difficult. Our quirks prod us toward directional innovation and away
from intersectional innovation, *even if the risks in both approaches are*
the same.

But if we understand where our fear of failure comes from, we can
fight it. To do that, we will delve deep into the subtle and strange world
of human psychological inconsistencies. We will learn what traps us
within a field, and what we have to do to achieve a balanced view of the
risks involved.

Trap 1: If Things Are Going Well, We Stay Within a Field

Suppose, for a moment, that you were forced to make the following
choice:

You must either pay $3,000 or take a gamble with an 80 percent
risk of paying $4,000 and a 20 percent chance of paying nothing.[3]

Did you take the chance to possibly pay nothing? Most people do.
Ninety-two percent of respondents in an experiment said they would
gamble on paying $4,000 with a chance of not having to pay anything.
This seems curious considering the mathematical expected loss is
$3,200 (80% x $4,000 = $3,200) for the gamble, $200 more than the

guaranteed loss. These results refute a common belief that most of us avoid risk. In this type of situation, it turns out, the vast majority of people welcome it. But what happens if the question is inverted?

> *You will either be given $3,000 or have to take a gamble with an 80 percent chance of winning $4,000 and a 20 percent risk of getting nothing.*

In an intriguing reversal of values, most people choose not to take the gamble under these circumstances. Instead, 80 percent of respondents opted for the safe cash. Suddenly, we have gone from welcoming risks to avoiding them, even though the mathematical expected gain for the gamble is $3,200. How can we explain the reversal?

The two psychologists who conducted these studies, Daniel Kahneman and Amos Tversky, developed a theory they called *prospect theory* to explain their observations.[4] Prospect theory suggests that it is not so much that we hate uncertainty, but rather that we fear losing. It is not that easy to see how things in our life could instantly get better—but it is easy to see how they could quickly get far worse.[5] That's why we are willing to gamble to avoid certain loss (of money), as in the example above. Loss is more vivid than gain. It is easier to imagine. It is more painful. And we fear it.

Our irrational reactions to possible loss can easily be observed outside of a laboratory. The obvious example is the stock market. When a stock has a run-up in value we are likely to sell in order to secure a gain. But when the stock drops in value we are more likely to hold on, hoping that the trend will reverse. This is not just true for amateur investors; it also holds true for professionals. The problem here is that if we take chances only when we have something to lose and play it safe when we have something to gain, we will be losing in the long run.

I have a friend, Martin, who is an experienced poker player. He is also (usually) a winning poker player. "Okay," he tells me when I ask him how he does it. "It has less to do with reading the other players around the table or figuring out if they are bluffing or not. It may help, but it's not the key to winning. The key to winning is to stay disciplined. Make sure to bet big when you have good cards and stay low when you don't. Because it is all about winning more when you win and losing less when you lose." Martin has observed countless times how people make riskier bets when things are going poorly, but lock in a win early when things are going well. They are the people who end up losing money in the long run.

This explains why we tend to stay within our own field when things are going pretty well instead of venturing toward the Intersection. Most of us would rather coast than risk losing what we have. It is comfortable and often very prudent to move forward in small, controlled steps, making sure to reap the gains we know we can get. A scientist who has done well in her field can maintain her status by pushing the field forward incrementally. A company that is leading the way within a particular market or product will stay in that market as long as they can. After all, we'd feel like idiots if we squandered a sure gain. The upshot of all this is that we become reluctant to try out intersectional ideas because taking risks would jeopardize our current level of status and security.

This behavior stands in stark contrast to our actions when things are not going well. This is usually when we take the really big chances, when we are willing to try something new. For instance, a company might test a radical strategy and go for broke if nothing else is working. Individuals who fear they will be laid off may see it as an opportunity to test out a new idea. The same holds true for scientists who work in a field where funding is drying up. They are left with no choice but to enter a new field, sometimes making remarkable intersectional discoveries.

The problem with all of this is that if we are willing to take risks and pursue intersections only when we are doing poorly, we'll hurt our overall chances of success. This is the point when we tend to be short on resources, contacts, credibility, and time. This is when we have the lowest chance of executing past our failures. Instead, we should try to innovate, to take more chances, when things are going well. When we are succeeding we have the highest chance to withstand failure, a necessary step for realizing intersectional ideas.

Both Richard Branson and Howard Berke managed to avoid this behavioral trap. Branson started his airline at a time when Virgin Music was doing quite well. Everybody else in his company was thinking about how to push further within the field of music. Branson, on the other hand, saw an opportunity to try something really different, and he took it. Berke is the same way. If his company-of-the-moment is doing well, he will leave it and search for another intersectional idea. He realizes that this is his best chance to switch fields and live to tell about it.

Trap 2: Time Spent in a Field Becomes a Reason to Stay in the Field

Imagine that you've invested $10 million developing a new solar energy technology but have nothing to show for it. Would you invest more?

Maybe. Then again, maybe not. It depends, but the answer has nothing to do with the $10 million you have already invested. Any standard economics textbook will tell you that the decision to invest more time or money should be based on what's going to happen in the future. Economists call the money spent "sunk cost" because capital invested is gone and cannot be taken back. It is out of the equation, and all that matters now is what the future can provide. You may, for instance, have learned from your initial attempt that the technology doesn't work, in which case investing more makes little sense. Or you might have learned that the technology not only works, but is also desperately needed, in which case it makes sense to invest more. Either way, the fact that you've already sunk $10 million into the project is essentially meaningless.

The same principle applies to time spent in a field. If you have spent years within a field, that fact alone can convince you to stay put even if it's a lost cause. A friend might complain to you that he no longer likes his job, saying; "I've invested so much in this career that it just isn't worth it to break off at this point." Like the fear of loss, this is another risk-related example of an emotional entanglement. If we have invested heavily we figure we should keep on investing. But the truth is that regardless of whether we're talking about time or money, both are sunk costs. Since they cannot be retrieved, only the future matters. This emotional trap creates a substantial barrier to stepping into the Intersection, even if we've found remarkable ideas worthy of pursuit. It is also a fairly common trap, difficult to avoid. But simply being aware of it makes it easier to overcome, and helps us choose to move on. Both Berke and Branson exhibit strong emotions regarding their creative efforts. Yet these emotions are always moving them forward. What they have done in the past does not by itself become a criterion for what they should do in the future.

Trap 3: We View Risks at the Intersection from a Directional Perspective

Imagine that a terrible disease has broken out in your community and you are the health care strategist in charge of taking action. It is believed that 600 lives are at stake and you can choose between one of two vaccines. The first vaccine will definitely save 200 lives. The second vaccine is experimental and has an uncertain outcome. It offers a 33 percent chance that all 600 people will be saved, but a 67 percent chance that no one will be saved. What would you do?

Kahneman and Tversky found in their studies that most people choose to save the 200 lives. Now, imagine that you instead had to choose between the following two options: With the first vaccine, 400 of the 600 people will die. The second vaccine is experimental and has an uncertain outcome. It provides a 33 percent chance that no one will die, and a 67 percent chance that everyone will die. What would you do?

In this version, 78 percent of respondents said they would try the experimental vaccine. This is rather interesting considering that the two situations described are exactly the same; they were just expressed differently. In the first case the situation was framed as saving 200 lives, in the second as letting 400 people die. People's risk-taking behavior changed significantly depending upon how they read the words "save" and "die."

This famous experiment suggests that people are deeply influenced by how a particular problem is framed. Given different presentations, the same situation may be seen as both risky and safe even by the same person. When researchers pointed out the discrepancy to their subjects, Kahneman and Tversky noted, they were "typically perplexed. Even after rereading the problem they still wish to be risk-averse in the 'lives saved' version." Their research tells us that it makes sense to view any risky situation from many different perspectives.

Unfortunately, we often view risks at the Intersection from only one perspective, the one we gained while working in a more directional field. The risks within established fields are better defined. In these cases we understand what is at stake. But if we maintain these frames of reference while evaluating ideas at the Intersection, we will always reach the conclusion that the uncertainty is too great. Even the best intersectional idea can seem too risky if we view it from the wrong frame of reference. And we will be convinced to stay away from it. So how can we move past this?

The people I have met all managed to do it by shifting their perspective regarding intersectional risks. Howard Berke, for instance, is focused on *learning*. "I was born poor and will probably die poor, but in between I will have learned a lot and had a great time," he says. Berke wants to understand how new industries work and be at the forefront of new fields. Through this lens, his pursuit of intersectional ideas seems far less risky, even downright safe. Richard Branson values the *fun* of doing something different. It is his "most important business criterion." Today his strategy is to intersect Virgin's businesses with a variety of different fields and industries. He is not sure up front whether something is going to work, but if it involves something fun and intrinsically motivating, he is taking a very small risk of being bored.

Marcus Samuelsson has yet another perspective; "Doing something differently, taking the chance to do something no one has done before; *it's my only chance to make it*," he says. If he had stuck within an established field of cooking it would have been much more difficult for him to innovate. John Seely Brown points out that it is virtually

impossible not to come out of the Intersection, "that white space between disciplines," without a *vastly expanded set of future opportunities*, regardless of whether the actual idea was a success or failure.[6]
Thus, exploring intersectional ideas will always yield downstream benefits—making it a fairly low-risk proposition.

Here is another view of such a perspective switch: The Intersection unleashes great creative powers through the explosion of concept
combinations. If your goal is groundbreaking innovation, this explosion
represents a gold mine of ideas. It would be crazy not to start digging.

Acknowledge Fear and Risks

T HERE IS A PASSAGE in Branson's autobiography that cannot
possibly leave any reader untouched. When this passage begins
Branson is planning to cross the Pacific Ocean, from Japan to America,
in a hot-air balloon together with his colleague Per.

> *Per told me his worst fear when it was too late. We were on the
> plane on our way to Japan when he confessed that he had been
> unable to test the capsule in a pressure chamber to be 100 percent
> sure that it would survive at 40,000 feet. If a window blew out at
> that height, we would have between seven and eight seconds to
> put on our oxygen masks.*
>
> *"We'll need to keep them handy," Per said in his usual under-
> stated way. "And, of course, if the other person is asleep, then it'll
> be necessary to put the mask on and get it going in three seconds
> and then put on the other person's in three seconds, allowing for
> two seconds for a fumble."*[7]

Why would Branson put himself through this? The guy *owns* an airline.
Why is he flying across the Pacific in a balloon? It just does not seem
normal. In fact, psychoanalysts less than a hundred years ago came to

the conclusion that risk seekers, like Branson, were lunatics. Branson understandably has a different take on it. "I never think I am going to die by accident, but if I die then all I can say is that I was wrong, and the hardened realists who kept their feet on the ground were right—but at least I tried." This may seem like a cavalier attitude, but from that remark and the preceding passage, we can unearth two important tools for overcoming fear: The first is acknowledging fear and the second is admitting that one can fail.

When you get scared, the human body makes it abundantly clear. The physiological reactions to fear are well documented. When you recognize a dangerous situation your blood pressure rises, your heart rate shoots up, your mouth dries out, and your palms become sweaty. Blood drains from unimportant areas such as your stomach and is redirected to the muscles, making you feel nervous and giving you butterflies. You become hyper-alert, supercharged on adrenaline, every sense heightened and ready for whatever is coming.

Today the nature of dangers has changed, yet the response has not. Instead of a lion circling to attack you and your children, you could be sitting in your bedroom late at night when you hear a strange noise from the kitchen. Your body will react in milliseconds. This hard-wired response kicks in even when physical danger is not imminent. It can be switched on by making a key decision such as leaving your job, changing your firm's strategy, selling your ideas to a tough audience—or even by a simple thing like making a sales call. Our understanding of risks ultimately boils down to this feeling of fear. How do we control it?

The most effective way to combat fear is to acknowledge it. When NASA studied the effects of space travel on its astronauts, it noticed that some consistently suffered from motion or stress sickness, what NASA deemed to be manifestations of fear. Other astronauts didn't. The space agency concluded that the major difference between the two groups was that the second group had acknowledged in advance that they were going to be afraid, whereas the first group had not.[8]

What does it mean to acknowledge fear? For starters, you have to come to terms with what is at stake and admit that you might lose it.

Often this means that you must be comfortable enough to know that if everything is lost, you can still move on. "If you are betting the ranch, you had better be able to pick up the pieces if things turn out differently from what you expect," says Peter Bernstein.[9] That is different from simply accepting a certain risk of failure. If someone gives you a fifty-fifty shot at winning $300 or losing $150, you may take the gamble. But if the same person gave you the same odds at winning $500,000 or losing your house valued at $250,000, you would probably walk away. Although the risk of failure is the same, most of us couldn't cope with losing our home.

We cannot always escape our fears, but we can manage them. By accepting our fears, by acknowledging that we can fail, and by becoming comfortable with what happens if we do, we can much more effectively move toward realizing our ideas at the Intersection. In the words of Mark Twain: "Courage is resistance to fear, mastery of fear, not absence of fear."

The Courage to Move On

WE ALL HAVE EMOTIONAL QUIRKS that push us toward directional innovation and away from the Intersection. By being aware of this we can, in a subtle way, counteract this push and face our fears. That is what Branson and Berke do, and we can, too.

But this is difficult advice to follow even if you are aware of your fears. Consider something Larry Susskind, a professor at MIT and a visiting professor at Harvard Law School, told me one day.[10] Although he never pursued a law degree, Susskind specializes in negotiations and has mediated large-scale disputes all over the world in most types of industries. During his career Susskind has zigzagged through a plethora of fields. He majored in English, earned a Ph.D. in urban planning, and then served as external director for an environmental consulting firm, as a planning consultant, negotiation advisor, and policy analyst, working in China, Spain, Japan, and Israel. Through all of that he has become one of the most innovative leaders in conflict resolution. So I asked Susskind one morning if he believes his insights would have been possible if he had stuck with one established field and shied away from the Intersection.

"Well, no," he admitted, leaning back in his chair. "I do believe in this stuff," he says. "I really do. The greatest risk is not taking one." He hesitates for a second before going on. "But what happens when you have to give advice to others that you care about, like your kids? I'm not sure what to tell my kids. Do I tell them that all you have to do is take chances, not to specialize, not to focus? I know that specializing will do well for them in life. So I just don't know; I don't know what to tell them."

Neither do I. It depends. Susskind pursued the intersection of disciplines and cultures, and he broke new ground because of it. But will his children eventually have that same propensity for discovery? Maybe they will; maybe they won't. Maybe they'll go on to improve the world within well-established fields and perhaps they'll do that better than anyone before them. Like all of us, his kids will eventually have to make up their own minds. But what I do know is that if they wish to break new ground, stepping into the Intersection will give them the most opportunities to do so—today, more than ever. It is the best chance we have to change the world. We should take it.

fifteen

Step into the Intersection . . .

AND CREATE THE MEDICI EFFECT

THERE IS A MYTH about glass. The myth says that glass is a liquid that in ordinary temperatures just looks like solid material. But if we had the patience, and the ability to watch it closely, we would see that glass is actually slowly flowing, like very thick, very sticky glue. The myth probably originated from the observation that old church windows appear to be thicker at the bottom than at the top. It has become a legend, something even knowledgeable teachers and parents tell children. But it is not true.

Egon Orowan at MIT commented on the phenomenon: "Half the pieces in a window are thicker at the bottom. But," he added, "the other half are thicker at the top."[1] If glass flows at the speed this myth would have us believe, glass vials in Egyptian tombs would be puddles today. But it is easy to see how such an urban legend could catch on. There is something mysterious about glass. Glass is clear; glass can be shaped and reshaped through heat; glass can both shatter light and direct it; and glass can help us see things from very, very far away. Yes, glass has amazing properties, but most of them are by design. They are the result of hundreds of years of innovations.

The observed "flow" in old church windows is a result of the uneven manufacturing process of the time. Today glass design sees new breakthroughs faster than ever before. Consider, for instance, the wonders of something called optical fibers. Optical fibers are pulled glass, and they are as thin as a single strand of hair. The fibers can stretch across vast distances and in recent years have been laid across enormous swaths of land in every continent in the world. A laser at one end can switch on and off 10 billion times per second, sending enormous amounts of information down a single strand of glass and across an entire continent. One hairline-thick strand of glass can transmit 5 million simultaneous phone conversations.

One company has had more innovations involving glass than any other in the world. This company, Corning Inc., has a laboratory in Sullivan Park, New York, where researchers spend their time hunting for intersections between various ways of using glass and the fundamentals of physics, mathematics, and chemistry. Every day at Corning more than a hundred types of glass experiments are conducted in Sullivan Park. This building houses what is arguably one of the most innovative research groups in the United States.

Corning has a long history of innovation. Ever since it was founded, over 150 years ago, Corning has affected our lives in noticeable ways. It created and manufactured glass bulbs for Thomas Edison's lights. It made the color tubes for virtually every single television set in the United States. It made the glass for the thermometer and the glass for LCD screens for electronic displays, and it owned both of those markets. Perhaps Corning's most widespread innovation is the glass casserole dish (sold by Pyrex) that can be taken directly from the freezer to the oven without danger of cracking. And, of course, they invented optical fibers, which became the basis for the entire telecommunications boom. Corning is, in other words, one of those rare companies that has managed to stay on the cutting edge for more than a hundred years.[2]

How did the company get this way? If anyone has the answer, it is Lina Echeverria, the head of the glass research group. Her department is the most watched inside the company, and the one Corning depends on for future success. Echeverria is Colombian; she came to the United States after spending a couple of years doing research in Germany. What you notice about her right away is that she has an incredibly strong personality. She is very upbeat, full of energy and armed with stories.

When she talks about her early days, studying lava rocks on an island off the coast of Colombia, her voice is filled with excitement.[3] "It was Gorgona Island, where Pizarro lost 90 percent of his men from snake bites. There were poisonous snakes everywhere. My research partner left after only a couple of days; he just couldn't take it," she says, but the snakes were not the only unusual research condition. "Besides the snakes the entire island had been converted into a prison for criminals—the worst kind. My guide was one of the prisoners since he knew the island the best." She was also accompanied by an armed guard, who kept an eye on the prisoner and the snakes. "Yeah, it was pretty exciting," she adds.

Echeverria has passion for almost everything she undertakes. She seems intent upon making sure everyone else does, as well. "I want the researchers [at Corning] to have the creativity of van Gogh, but lead the life of Michelangelo," she says. I ask her what she tells her researchers to encourage them to step into the unknown and to innovate. "I tell them to follow their hearts," she says. "Follow your heart. Do something you are interested in, do something you can get energized about. That is where passion comes from. And," she says, "creativity comes from passion."

She encourages people to interact, share, and collaborate in order to create or join projects they are excited about. Echeverria even created a special "creativity room," where people can talk about whatever is on their mind, to encourage cross-fertilization of ideas.[4] "You create a group of people that act as sounding boards," she says. Often people have to be guided to find these connections. Echeverria considers her ability to "get the right person on the right project" the most important part of her job.

She tells me a story about Doug Allen, a theoretical physicist at Corning who mostly sat in a corner of the lab and worked on cutting-edge research in quantum mechanics. Echeverria had a hunch that Allen was more social than he let on, so she asked him to join a team working on an actual product, something he had never done. By bringing his knowledge into a product group, Allen's work suddenly was able to have a greater impact on Corning's bottom line than anything he had done in the past eight years.

This, then, is one reason why Corning has managed to stay on the cutting edge for so long. The company keeps its people passionate about searching for new intersections. It is a strategy that will serve them well in the future, as the number of intersections in the world continues to rise. We have seen dozens of examples of this throughout the book. We have met teams and individuals who have searched for, and found, intersections between disciplines, cultures, concepts, and domains. Once there, they had the opportunity to innovate as never before, creating the Medici Effect. That has been the message of this book. What is left to say?

A few things. I want to leave you with three key points. They represent the last chance for me to inspire you to seek out the Intersection. I'm going to take it.

The Future Lies at the Intersection: Find Your Way There

I N T H I S B O O K I have talked about games and navigational systems, about food and solar power firms. Because of the three driving forces—the movement of people, the convergence of science, and the leap of computation—these areas and countless others are becoming increasingly interconnected.

For instance, one day I was talking with Michael Dukakis, the former governor of Massachusetts. Besides running against George Bush Sr. in the presidential election of 1988, he is also known for the creation of a nation-leading health care system in his state. He pointed out how interdisciplinary the problems of health care delivery have become. "You have dissatisfied and angry doctors, HMOs, nurses, insurance companies, unions and employers, drug companies, and governments, and yet we are delivering health care at a cost double that of the rest of the Western world." He was convinced that the solution could be found at the intersection of these fields.

The same can be said for the growing battle against global terrorism, where independently financed and globally mobile small terrorist groups defy standard defense tactics. The newly launched U.S. Department of Homeland Security attempts to fight this war by merging twenty-two different federal entities, including the Immigration and Naturalization Service, the Coast Guard, and the Secret Service. If these institutions can break down the barriers that exist between them, we may see innovative approaches to tackling the threat. Or consider global warming; scientists including chemists, oceanographers, ecologists, and geologists all work together to understand it and predict its effects.

The solutions to terrorist threats, health care crises, and environmental problems are multifaceted and do not easily fit into distinct fields. But neither do solutions to less dramatic challenges, such as better fashion design, product innovation, and animated movies. In every arena, whether in the sciences or the humanities, business or politics, there is a growing need to connect and combine concepts from disparate fields. That is how we will find new opportunities, surmount new challenges, and gain new insights. That is the way we will create our future. The future lies at the Intersection, and if you wish to help create it, find your way there.

Expect the Unexpected, Because Intersections Are Everywhere

EVERYTHING CONNECTS in one way or another. The trick is seeing how things connect and then knowing how to use those connections. This book has offered a lot of ways for doing just that.

Much of the advice boils down to this: *Expect the unexpected.* If you do you, will start seeing the world from new perspectives. Suddenly you will find intersections everywhere. Random conversations, meetings, or projects will begin to flow together in strange, but intriguing ways. Seemingly unrelated concepts will connect in ways you did not think were possible.

Who would have guessed that when Richard Garfield went to Multnomah Falls he would get an idea that connected collectibles with card games, and that such a connection would forever change the world of games? I don't think anyone could have predicted that Deborah Prothrow-Stith would find a connection between violence prevention and health care. Today this intersection of fields seems obvious, but it wasn't always. She spotted it at 3 A.M. one cold January night in a Boston ER. Could anyone have guessed that when telecom engineer Eric Bonabeau met insect ecologist Guy Theraulaz, it would ultimately help truck drivers find their way around the Swiss Alps? Probably not, although one could have guessed that *something* unexpected would come out of their discussions.

This is the nature of intersectional ideas. If you let them happen, they will. You may not know exactly when or exactly where. But when one of them hits you—be ready for it. Prepare to be surprised. Expect the unexpected.

There Is Logic to the Intersection, But the Logic Is Not Obvious

THE UNEXPECTED NATURE of the Intersection makes it a place of uncertainty. It is unknown territory where past knowledge and experiences are poor guides. This turns our standard way of looking at the world on its head. We are used to picking a destination and then heading toward it. That is common sense; that is logical. But the Intersection is a place where we must leave many of our preconceived notions behind. There *is* logic to intersectional ideas, but the logic is not obvious.

Obvious logic, for instance, tells us that it makes sense to prepare and budget a detailed plan of execution while pursuing directional ideas. What is not so obvious is that doing so at the Intersection can lead to failure. Obvious logic tells us to develop detailed, clear reward structures while pursuing directional ideas. What is not so obvious is that this will be self-defeating at the Intersection. Seemingly obvious logic tells us that having more resources should reduce the risk of failure at the Intersection. What is not so obvious is that the more resources we have, the more we will use—and thus the risk of failure remains the same. We may also find it strange that we do not have a better chance of achieving groundbreaking innovation by specializing in a field. But if we step into the Intersection, we can go from a mere 2,400 available concept combinations to almost six million—how do you compete with that?

Not everything that happens at the intersection of cultures, disciplines, concepts, and domains will be obvious. But when you understand the rules of the Intersection, it will begin to make sense.

Take the Leap

TODAY THERE ARE more reasons than ever to seek out the Intersection. Disciplines and cultures are connecting faster, more often, and in more places than ever before. In this book we looked at people who are taking advantage of these forces as they explore the intersection of fields. We will see others, many others, like them.

We can all create the Medici Effect because we can all get to the Intersection. The advantage goes to those with an open mind and the willingness to reach beyond their field of expertise. It goes to people who can break down barriers and stay motivated through failures. But we can all do that.

Most of us have a desire to connect ideas and concepts from our disparate backgrounds. So why not actively seek out these connections? While writing this book I met a vast number of people who were working in one area they find interesting, but at the same time expressed marked interest in another. Someone working in the nonprofit world might want to use their ideas for for-profit practices; another person might wish to link two different cultures. "If I could just find a way to connect these fields, bring the pieces together," they say, "then I could come up something exciting, something new." Well, they are right.

In our world it actually makes sense to combine sea urchins with lollipops, guitar riffs with harp solos, and music records with airlines. In our world it makes sense for spiders and goat milk to have something in common or for a person to launch a solar cell company one day and a cookie company the next. Like the creators of fifteenth-century Florence, this is how we break new ground; this is how we innovate.

The world is, in some ways, like a giant Peter's Café, the place where sailors from every port on the planet stop for a beer, a conversation, and a chance to mix and combine ideas. The world is connected and there is a place where those connections are made—a place called the Intersection.

All we have to do is find it . . . and dare to step in.

Notes

Introduction

1. Donald McNeil, "Termite Mounds Inspire Design for African Office Complex," *New York Times*, 13 February 1997.

2. Quoted from "Ant Hill," an article by Lindsay Johnston that describes Eastgate, among other buildings. See <www.fourhorizons.com.au/lindsay/ar_articles/ar_74.pdf>.

3. For a great portrait of George Soros's life, philosophy, and accomplishments, read Michael Kaufman, *Soros: The Life and Times of a Messianic Billionaire* (New York: Alfred A. Knopf, 2002). Also, for a good account of how George Soros integrated philosophy into his trading strategies, see George Soros, *The Alchemy of Finance: Reading the Mind of the Market* (New York: Simon & Schuster, 1987).

Chapter 1

1. You can actually view this game at the Web site: <donoghue.neuro.brown.edu/multimedia.php>.

2. Mijail Serruya et al., "Instant Neural Control of a Movement Signal," *Nature*, 14 March 2002.

3. Emily G. Boutilier, "Monkey Mind: The Most-Reported Brown Science Story Ever?" *Brown Alumni Monthly*, May/June 2002.

4. Leon Cooper, interview by author, Brown University, Providence, RI, August 2001.

5. Mark Williams, "Profile: For Leon Cooper, Biology and Technology Are Merging," *Red Herring*, January 2000.

6. There are many others. Leon Cooper's interest in combining fields can also

be found in his teaching at Brown University. He has, for instance, developed an introductory physics course that also provides credit for coursework in theater arts!

7. The quotes in this paragraph come from the following article: Scott Turner, "Researchers Demonstrate Direct, Real-Time Brain Control of Cursor," *George Street Journal*, 15 March 2002. The remaining information from Serruya in this chapter comes from a personal interview and correspondence during the summer and fall of 2002.

8. John Donoghue, interview by author, August and September 2002.

9. Teresa Amabile, *Creativity in Context* (Boulder: Westview Press, 1996), 35–37.

10. Robert Sutton, *Weird Ideas That Work* (New York: Free Press, 2002).

11. Sarnoff Mednick emphasizes this point in his groundbreaking paper, "The Associative Basis of the Creative Process," *Psychological Review* 69, no. 3 (1962): 220–232.

12. Although there have been numerous attempts to find objective criteria for creativity and innovation, such approaches suffer from several drawbacks. See Teresa Amabile's book *Creativity in Context* for a detailed discussion on this. Mihaly Csikszentmihalyi is a strong advocate of using society as the measure for a creative idea. His arguments are outlined most comprehensively in *Creativity: Flow and the Psychology of Discovery and Invention* (New York: Harper Perennial, 1996).

13. Csikszentmihalyi, *Creativity*, 23.

14. Richard Dawkins, *The Selfish Gene* (Oxford, UK: Oxford University Press, 1976).

15. Ibid., 192.

16. Most of the academic information available for this field can be found online. See, for instance, <http://jom-emit.cfpm.org/> for one of the more prominent journals.

17. See, for instance, Robert Wright's book *Nonzero: The Logic of Human Destiny* (New York: Vintage Books, 2001).

18. The person who has taken this particular idea farthest is Seth Godin. Read his book *Unleashing the Ideavirus* (New York: Hyperion, 2001) for a comprehensive description of how idea viruses can be used for marketing purposes. To read about the Hush Puppy, see Malcolm Gladwell, *The Tipping Point* (New York: Little Brown, 2000).

19. Probably the most comprehensive paper discussing this balance is James G. March, "Exploration and Exploitation in Organizational Learning," *Organizational Science* 2, no. 1 (1991): 71–87.

Chapter 2

1. Thomas Friedman, *The Lexus and the Olive Tree* (New York: Anchor House, 2000).

2. There are many different sources for this story, with no definitive version. The main source for this recounting comes from <www.allthingscherokee.com>.

3. "Half a Billion Americans," *The Economist*, 22 August 2002.

4. Organization for Economic Co-operation and Development, *Trends in International Migration: Continuous Reporting System on Migration* (Paris: OECD, 2001).

5. Peter Drucker, "The Next Society: A Survey of the Near Future," *The Economist*, 3 November 2001.

6. "Shakira," *Contemporary Musicians* 33, Gale Group, 2002, <http://www.galegroup.com/free_resources/chh/bio/shakira.htm>.

7. John Leland and Veronica Chambers, "Generation Ñ," *Newsweek*, 12 July 1999.

8. Yvonne Chu, interview by author, New York, April 2003.

9. Jim Robbins, "Second Nature," *Smithsonian*, July 2002.

10. Perhaps the most adamant proponent for this way of viewing the world is John Horgan, author of *The End of Science* (New York: Addison-Wesley, 1996).

11. "Seashells Yield Tough Secrets," *Christian Science Monitor*, 19 August 1997.

12. Alan Leshner, interview by author, August 2003.

13. George Cowan, interview by author, March 2002.

14. Robert Hagstrom, *Investing: The Last Liberal Art* (New York: Texere, 2000).

15. Pixar Annual Report, <http://www.pixar.com/companyinfo/investors/annual reports/1996/>.

16. There are numerous accounts of this trend. Check out this Web site at Mouse Planet for a revealing story by David Koenig on how Disney has shut down its traditional animation efforts: <http://www.mouseplanet.com/david/dk030814.htm>.

17. Mark Tracy, interviews by author, November 2002 and March 2003.

Chapter 3

1. Ruth Reichl, "A New Chef Expands on the Swedish Tradition of Balancing Salty and Sweet," *New York Times*, 29 September 1995.

2. The information about Aquavit and Marcus Samuelsson was provided through interviews with Jennie Andersson, PR manager for Aquavit, and from Aquavit's press kit.

3. Milford Prewitt, "Aquavit," *Nation's Restaurant News*, 22 May 2000.

4. Marcus Samuelsson, interviews by author, February and April 2002 and February 2003.

5. This example is adapted from one in Michael Michalko's book *Cracking Creativity* (Berkeley: Ten Speed Press, 2001).

6. An excellent discussion on this topic can be found in Sarnoff Mednick, "The Associative Basis of the Creative Process," *Psychological Review* 69, no. 3 (1962): 220–232.

7. J. P. Guilford, *The Nature of Human Intelligence* (New York: McGraw-Hill, 1967).

8. Test developed by H. J. Eysenck, *Genius: The Natural History of Creativity* (Cambridge: Cambridge University Press, 1995).

9. Dean Simonton, *Origins of Genius* (New York: Oxford University Press, 1999).

10. The information for how Darwin came upon the idea of evolution was found at <http://www.aboutdarwin.com> and <http://www.pbs.org/wgbh/evolution/darwin/diary>.

Chapter 4

1. Donald Campbell, "Blind Variation and Selective Retention in Creative Thought as in Other Knowledge Processes," *Psychological Review* 67, no. 6 (1960): 380–400.

2. Dean Simonton, *Origins of Genius* (New York: Oxford University Press, 1999).

3. A. J. Lopez, G. B. Esquivel, and J. C. Houtz, "The Creative Skills of Culturally and Linguistically Diverse Gifted Students," *Creativity Research Journal* 6 (1993): 401–412; and D. C. Carringer, "Creative Thinking Abilities in Mexican Youth," *Journal of Cross-Cultural Psychology* 5 (1974): 492–504.

4. Marcus Samuelsson, interview by author, February 2002.

5. Paul Maeder, interview by author, December 2002.

6. Robert Sternberg, *Successful Intelligence: How Practical and Creative Intelligence Determine Success in Life* (New York: Plume, 1997).

7. Thomas Kuhn, *The Structure of Scientific Revolutions* (Chicago: The University of Chicago Press, 1962).

8. Simonton, *Origins of Genius*.

9. Paul Israel, *Edison: A Life of Invention* (New York: John Wiley & Sons, 1998), 12.

10. Simonton, *Origins of Genius*.

11. Two great books that go into detail on this topic are the classic by John Adams, *Conceptual Blockbusting* (New York: Perseus Books, 1974), and Michael Michalko, *Cracking Creativity* (Berkeley: Ten Speed Press, 2001).

12. The history about codes and code breakers was gleaned from Simon Singh's outstanding work *The Code Book* (New York: Anchor Books, 2000).

13. Michael Michalko, *Cracking Creativity* (Berkeley: Ten Speed Press, 2001), 177.

14. The exercise of becoming a flower is from Michael Schrage, *Serious Play* (Boston: Harvard Business School Press, 2000).

15. Michalko, *Cracking Creativity*.

16. Markus Ahman, interview by author at the Institute of Water and Air in Stockholm, Sweden, October 2002. Also see the informative book by A. Denny Ellerman et al., *Markets for Clean Air* (Cambridge: Cambridge University Press, 2000).

Chapter 5

1. "Magic: The Gathering—Playing with Fantasy in a Postcinematic Form," in Kurt Lancaster, *Warlocks and Warpdrive: Contemporary Fantasy Entertainments with Interactive and Virtual Environments* (Jefferson, NC: McFarland, 1999), 47–58.

2. The market information regarding Magic is from correspondence with Tod Steward, PR Group Manager at Wizards of the Coast, February 2002.

3. Richard Garfield, interview by author, February 2002.

4. N. R. Maier, "Reasoning in Humans: II. The Solution of a Problem and Its Appearance in Consciousness," *Journal of Comparative and Physiological Psychology* 12 (1931): 181–194.

5. Arthur Koestler, *The Act of Creation* (London: Penguin Arkana, 1964).

6. Arthur Koestler, "The Three Domains of Creativity," in D. Dutton and M. Krausz, eds., *The Concept of Creativity in Science and Art* (The Hague, Netherlands: Martinus Nijkoff Publishers, 1981).

7. Robert Johnson, "The Market Nobody Wanted," *Fortune*, 1 August 2002.

8. Sarnoff Mednick, "The Associative Basis of the Creative Process," *Psychological Review* 69, no. 3 (1962): 220–232.

9. Malcolm Gladwell, interview by author, March 2002.

10. Dean Simonton has done a very good job of classifying the different types of serendipitous experiences. See Dean Simonton, *Origins of Genius* (New York: Oxford University Press, 1999).

11. This story is repeated often in the creativity literature. A great summary of Pasteur's discovery can by found in Arthur Koestler's *The Act of Creation*.

Chapter 6

1. A good short summary of theories on what killed off the dinosaurs can be found at <http://www.ucmp.berkeley.edu/diapsids/extinction.html>.

2. Luis Alvarez worked together with his son Walter Alvarez, who was a geologist.

3. This strategy is supported by a number of researchers in creativity. Dean K. Simonton suggests it in *Origins of Genius* (New York: Oxford University Press, 1999). He also makes clear that sometimes a switch to another field leads to new insights, but other times it doesn't. The difference can be found in one's ability to break down the associative barriers and execute past failures.

4. Edward DeBono, *New Think: The Use of Lateral Thinking in the Generation of New Ideas* (New York: Basic Books, 1968).

5. Orit Gadiesh, interview by author, March 2002.

6. Frank Herbert's history is a compilation of information from his son Brian Herbert's biography, *The Dreamer of Dune: The Autobiography of Frank Herbert* (New York: Tor Books, 2003) and from <http://www.arrakis.co.uk/herbert.html>.

7. When it was released, *Dune* was the first science fiction novel to win both the Hugo and Nebula fiction awards. It has since spawned five sequels, five prequels, two miniseries, and a feature-length movie.

8. Simonton, *Origins of Genius*.

9. How the team at Bletchely Park cracked the Enigma is described in Simon Singh's riveting work *The Code Book* (New York: Anchor Books, 1999). Throughout the war the team also had help from military teams that managed to capture Enigma machines from German submarines.

10. The idea that the United States gains part of its competitive advantage from its diversity has been suggested by many people. I heard it many times during my research, most notably from Paul Lawrence, professor at Harvard Business School, and from Luke Visconti, cofounder of the magazine *Diversity Inc*.

11. Steve Miller, interview by author, August 2003.

12. Donn Byrne, "Interpersonal Attraction and Attitude Similarity," *Journal of Abnormal and Social Psychology* 62 (1961): 713–715, and Donn Byrne, *The Attraction Paradigm* (New York: Academic Press, 1971).

13. There is a large volume of literature that illustrates the problems with

unstructured interviews. The following site contains a good summary of studies outlining these problems: <http://www.hr.com/HRCom/FormTemplates/DDI-SEWP01.pdf>.

14. Robert Sutton, *Weird Ideas That Work* (New York: Free Press, 2002).

15. Teresa Amabile, "How to Kill Creativity," *Harvard Business Review*, September–October, 1998.

16. John Donoghue, interview by author, August and September 2002.

17. See Dorothy Leonard and Susan Straus, "Putting Your Whole Company's Brain to Work," *Harvard Business Review*, July–August 1997.

18. Studies have verified this. See, for instance, Thomas B. Ward et al., "Creative Cognition," in *Handbook of Creativity*, ed. Robert Sternberg (Cambridge: Cambridge University Press, 1999).

19. Michael Michalko, *Cracking Creativity* (Berkeley: Ten Speed Press, 2001).

20. Richard Garfield, interview by author, February 2002.

Chapter 7

1. The story of how Håkan Lans got the idea for STDMA is from David Lagercrantz's, *Håkan Lans: Ett Svenekt Geni* (Stockholm: Bokförlaget DN, 2000).

2. Dean K. Simonton, *Origins of Genius* (New York: Oxford University Press, 1999).

3. Stanley Murray, "Virgin Billionaire: Polo.com Magazine's One-on-One with Sir Richard Branson," *Polo.com Magazine*, 2002.

4. For a list of Joyce Carrol Oates's complete works, see <http://www.usfca.edu/fac-staff/southerr/works.html>.

5. Håkan Lans, interviews by author, January 2002–January 2003.

6. Information about the STDMA system is available at <http://www.gpc.se>.

7. Robert Weisberg, "Creativity and Knowledge: A Challenge to Theories," in *Handbook of Creativity*, ed. Robert Sternberg (Cambridge: Cambridge University Press, 1999), provides an example of a similar line of logic.

8. Simonton, *Origins of Genius*.

9. Stuart Whitmore, "Driving Ambition," *Asiaweek*, 25 July 1999.

10. Simonton's research is summarized in *Origins of Genius*, but some facts in this section come from his earlier work, *Scientific Genius: A Psychology of Science* (Cambridge: Cambridge University Press, 1988).

11. Richard Branson gives an excellent account of this story in his autobiography *Losing My Virginity: How I Survived, Had Fun, and Made a Fortune Doing Business My Way* (New York: Times Business, 1998). Another great place to gain insight into this compelling story is the comprehensive Mike Oldfield fan site: <http://www.tubular.net>.

12. Andrew Valentine, "As Mike Oldfield Releases a New Album, an Astonishingly Frank Confession," *Daily Mail* (U.K.), 31 August 1998.

13. The quote from David Bedford comes from the songbook *Tubular Bells*, published by Wise Publications.

14. Mike Oldfield has sold around 45 million albums throughout his career. Each recording has had highly variable sales, but he has had major hits in every decade since the 1970s.

Chapter 8

1. James Adams, *Conceptual Blockbusting: A Guide to Better Ideas* (New York: Perseus Books, 1974).

2. J. P. Guilford, *The Nature of Human Intelligence* (New York: McGraw-Hill, 1967).

3. Michael Michalko, *Cracking Creativity* (Berkeley: Ten Speed Press, 2001).

4. Ibid.

5. Tom Kelley, *The Art of Innovation* (New York: Doubleday, 2001).

6. Alex Osborn, *Applied Imagination: Principles and Procedures of Creative Problem Solving* (New York: Scribner, 1957).

7. Michael Diehl and Wolfgang Stroebe, "Productivity Loss in Brainstorming Groups: Toward the Solution of a Riddle," *Journal of Personality and Social Psychology* 53, no. 3 (1987): 497–509.

8. John Rossiter and Gary Lilien, "New Brainstorming Principles," *Australian Journal of Management* 19, no. 1 (1994): 61–71.

9. The process for brainwriting was described to me by Teresa Amabile, professor at Harvard Business School.

10. For die-hard followers of the series, there are actually a number of Web sites that list the MacGyverisms of each episode. You can find a great one at <http://www.davidhales.net/macgyver/isms/>.

11. Teresa Amabile, "Creativity Under the Gun," *Harvard Business Review*, August 2002 (special issue).

12. Ibid.

13. Richard Garfield, interview by author, February 2002.

14. Håkan Lans, interviews by author, January 2002–January 2003.

Chapter 9

1. My sources for the story about Deborah Prothrow-Stith come from interviews with her through spring of 2002 and 2003 and from her two books: *Deadly Consequences* (New York: Harper Perennial, 1991) and *Murder Is No Accident: Understanding and Preventing Youth Violence in America* (New York: Wiley, 2004), which she cowrote with her colleague Howard Spivak.

Chapter 10

1. Richard Farson and Ralph Keyes, "The Failure-Tolerant Leader," *Harvard Business Review*, August 2000 (special edition).

2. Steve Kemper, *Code Name Ginger* (Boston: Harvard Business School Press, 2003).

3. These suggestions come from Robert Sutton, *Weird Ideas That Work* (New York: Free Press, 2002).

4. These statistics are from a presentation by Vertex's president Vicki Sato at a December 2002 pharma/biotech conference held by the consulting firm ISO Healthcare in New York City.

5. For a great story about Vertex Pharmaceuticals and its early strategy and

challenges, see Barry Werth, *The Billion Dollar Molecule: One Company's Quest for the Perfect Drug* (New York: Simon & Schuster, 1994).

6. Vicki Sato, interview by author, January 2003.

7. A demonstration of this can be seen at the Vertex Web site, <http://www.vrtx.com>.

8. Howard Gardner, *Creating Minds: An Anatomy of Creativity Seen Through the Lives of Freud, Einstein, Picasso, Stravinsky, Eliot, Graham, and Gandhi* (New York: Basic Books, 1993).

9. Pat Dillon, "The Next Small Thing," *Fast Company*, June/July 1998.

10. The Picasso reference comes from Gardner, *Creating Minds*.

11. Clayton Christensen, *The Innovator's Dilemma* (Boston: Harvard Business School Press), 1997.

12. There are two excellent books about the story of Boo.com. One is Ernst Malmsten (the founder), Erik Portanger, and Charles Drazin, *Boo Hoo: $135 million, 18 months . . . a Dot.com Story from Concept to Catastrophe* (London: Arrow, 2001). The other, written by investigative reporter Gunnar Lindstedt, is called simply *Boo.com* (Stockholm: Forum AB, 2001).

13. For a comprehensive text on intrinsic motivation and how it relates to creativity, see Teresa Amabile, *Creativity in Context* (Boulder, CO: Westview Press, 1996). Also see Mercer Mayer, *A Boy, a Dog, and a Frog* (New York: Dial Press, 1967).

14. Stephen King, quoted in Amabile, *Creativity in Context*.

15. Jim Collins, *Good to Great: Why Some Companies Make the Leap . . . and Others Don't* (New York: HarperCollins, 2002).

16. John Seely Brown, interview by author, October 2002.

17. Amabile, *Creativity in Context*.

18. Håkan Lans, interview by author, October 2002.

Chapter 11

1. Eric Bonabeau, interview by author, summer, 2003.

2. E. Bonabeau, M. Dorigo, and G. Theraulaz. "Inspiration for Optimization from Social Insect Behaviour," *Nature* 406 (2000): 39–42.

3. Derrick Story, "Swarm Intelligence: An Interview with Eric Bonabeau," The O'Reilly Network, 21 February 2003, <http://www.oreillynet.com/pub/a/p2p/2003/02/21>.

4. Julia Flynn, "British Telecom: Notes from the Ant Colony," *BusinessWeek*, 23 June 1997.

5. Eric Bonabeau et al. *Swarm Intelligence: From Natural to Artificial Systems* (New York: Oxford University Press, 1999).

6. Jeffery Rothfeder, "Expert Voices: Icosystem's Eric Bonabeau," *CIO Insight*, 11 June 2003.

7. P. Gaudiano et al., "Swarm Intelligence: A New C2 Paradigm with an Application to Control Swarms of UAVs" (paper presented at the 8th ICCRTS Command and Control Research and Technology Symposium, Washington, DC, June 16–20, 2003). The paper won the Gary F. Wheatley Best Paper Award.

8. Clayton Christensen, *The Innovator's Dilemma* (Boston: Harvard Business School Press, 1997).

9. Ibid., 179.

10. See Pixar's Web site for an overview of their history: <http://www.pixar.com>.

Chapter 12

1. <http://www.chopra.com/archives/aboutdeepak.htm>.

2. Deepak Chopra, interview by author, winter 2003.

3. See Chopra's Web site, <http://www.chopra.com>, for evidence of how extensive this enterprise has become.

4. To read about the JAMA controversy, see, for instance, Andrew Skolnick, "The Maharishi Caper: Or How to Hoodwink Top Medical Journals," *The Newsletter of the National Association of Science Writers*, fall 1991.

5. Linus Torvalds and David Diamond, *Just for Fun: The Accidental Revolutionary* (New York: HarperCollins, 2001).

6. Jim Kerstetter et al., "The Linux Uprising: How a Ragtag Band of Software Geeks Is Threatening Sun and Microsoft—and Turning the Computer World Upside Down," *BusinessWeek*, 3 March 2003.

7. Information about this exchange came from Torvalds's book *Just for Fun* and from a Web site created by Ragib Hasan, detailing the rise of Linux: <http://ragib.hypermart.net/linux/>.

Chapter 13

1. Richard Branson, *Losing My Virginity: How I Survived, Had Fun, and Made a Fortune Doing Business My Way* (New York: Times Business, 1998).

2. These issues are discussed at some length in Peter Bernstein's book *Against the Gods: The Remarkable Story of Risk* (New York: John Wiley & Sons), 1996. The quote from Keynes is also from this book.

3. European Commission, *Fostering Entrepreneurship in Europe: Priorities for the Future*, June 2000.

4. Howard Berke, interview by author, December 2002 and summer 2003.

5. Charles Smith, "History of Solar Energy: Revisiting Solar Power's Past," *Technology Review*, July 1995.

6. Gerald Wilde, interview by author, summer 2003.

7. This example and others about risk homeostasis in this chapter are from Gerald Wilde, *Target Risk* (Toronto: PDE Publications, 1994). The first edition of this book is available online at <http://pavlov.psyc.queensu.ca/target/>.

Chapter 14

1. Adam Lashinsky, "CEOs Under Fire: Now for the Hard Part," *Fortune*, 3 November 2002.

2. Peter Bernstein, *Against the Gods: The Remarkable Story of Risk* (New York: John Wiley & Sons), 1996.

3. All discussion in this chapter about Tversky and Kahneman's research is from

Peter Bernstein, *Against the Gods*, and Richmond Harbaugh, "Skill Reputation, Prospect Theory, and Regret Theory," working paper, Claremont Colleges, Claremont, CA, March 2002.

4. Kahneman was awarded the 2002 Nobel Prize in economics for these discoveries; Tversky had passed away at that point.

5. Bernstein, *Against the Gods*.

6. Marcus Samuelsson, interviews by author, February and April 2002 and February 2003; John Seely Brown, interview by author, October 2002.

7. Richard Branson, *Losing My Virginity: How I Survived, Had Fun, and Made a Fortune Doing Business My Way* (New York: Times Business, 1998), 213.

8. James McCormick, "How Overcoming Fear Can Boost Your Career," *CareerJournal.com*, <http://www.careerjournal.com/myc/climbing/19981105-mccormick.html>.

9. Bernstein, *Against the Gods*, 116.

10. Larry Susskind, interview by author, April 2002.

Chapter 15

1. Robert Brill, "No It Doesn't Flow," *The Corning Museum of Glass*, July 2000, <http://www.glassnotes.com/WindowPanes.html>.

2. Charles Fishman, "Creative Tension," *Fast Company*, November 2000.

3. Lina Echeverria, interview by author, October 2002.

4. The creativity room is also discussed in an article by Lee Bruno, "Putting Quirks to Work," *Red Herring*, 3 July 2002.

Index

AAAS. *See* American Association for the Advancement of Science
action
  changing plans and, 135
  failures and success and, 121–124
  ideas and, 114–116
  inaction and, 129
  planning and, 124–125
Adams, James, 106
Adkison, Peter, 61, 72
Adleman, Leonard, 55
advertising, and culture, 46
Allen, Doug, 186
Alvarez, Luis, 74
Amabile, Teresa, 6, 14, 112, 136, 140
American Association for the Advancement of Science (AAAS), 26–27
animation industry, 28–30, 149–151
Aquavit (restaurant), 35–38, 47–49, 60, 156. *See also* Samuelsson, Marcus
architectural design, 3–4
associative barriers
  assumption reversal and, 53–57
  cultural exposure and, 46–49
  defined, 38–41

effects of high vs. low barriers and, 41–43
how to break down, 45–60
learning differently and, 49–53, 92
perspective taking and, 57–60
assumptions, reversal of, 53–57

Bach, Johann Sebastian, 91
Bain & Company, 75–77
Bathia, Sabeer, 96
Bedford, David, 102
behavioral traps, avoidance of, 172–179
Berke, Howard, 164–166, 169, 176, 178
Bernstein, Peter, 172–173, 181
BET. *See* Black Entertainment Television
Black Entertainment Television (BET), 69
"blocking," 109–110
bogus-stranger technique, 80–81
Bonabeau, Eric, 143–146, 148, 152, 154, 188
boo.com (Internet company), 134
brain science. *See* mindreading experiment

brainstorming, 105–111
　issues in, 107–110
　rules for, 108, 109, 110–111
　solutions to issues in, 110–111
　targets for, 105–107
brainwriting, 111
Branson, Richard, 91, 98, 161–162, 176, 178, 179–180
Brown, John Seely, 140, 178–179
Brown University Brain Science Program, 11–13, 52, 105
Byrne, Donn, 80

Campbell, Donald, 47
Cargill, 30–32
Chopra, Deepak, 154–156, 157
Christensen, Clayton, 6, 132–133, 134, 147–149
Chu, Yvonne, 24
Cochrane, Peter, 144
Collins, Jim, 139–140
communication, 30–32, 83. See also teams
computational advances, 28–32
concept combinations. See also Intersection; Intersectional ideas
　creativity and, 67–69
　explosion of, 97–101
concepts, 16
convergence, in science, 25–28
Cooper, Leon, 12, 13, 191n6
Corning, Inc., 184–186
courage
　fear management and, 179–181
　at the Intersection, 164–166
　to move on, 181–182
　strategies for finding, 171–182
Covey, Stephen, 19
Cowan, George, 27–28
creativity
　associative barriers and, 40
　characteristics of, 14–15
　conceptual combination and, 67–69
　education and, 49–53
　effect of rewards on, 136–138, 139–140
　time pressure and, 111–114
creativity experiment, 66–67
"creativity room," 185
credibility, 128, 133, 135
Crichton, Michael, 145
Csikszentmihalyi, Mihaly, 15, 192n12
cultural diversity
　associative barriers and, 46–49
　globalization and, 22–24

Darwin, Charles, 42–43, 45, 53, 78, 95–96
Dawkins, Richard, 17–18
Diehl, Michael, 109
Diffie, William, 54–55
Dines, David, 31
directional vs. Intersectional ideas, 17–20
　motivation and, 138–141
　planning and, 124–126
　risks and, 163–164, 177–179
disk-drive market, 147–148
Disney. See Walt Disney Company
divergent thinking, 40
diversity
　challenges in, 83
　competitive advantage and, 195n10
　cultural, 22–24, 46–49
　occupational, 74–78
　perspectives and, 57–60
　teams and, 79–83
Donoghue, John, 13, 52, 83
Drucker, Peter, 23
Dubinsky, Donna, 131–132
Dukakis, Michael, 186–187
Dune (science fiction novel), 77–78

Eastgate office complex, 3–4
Echeverria, Lina, 185
Edison, Thomas, 52–53, 78, 91, 107
education. See learning
Einstein, Albert, 91

Eliot, T. S., 130–131
emotions, and risk, 172–173
encryption, 54–55, 79
environmental management, 58, 187
evaluation
    brainstorming and, 109
    time for, 111–114
"evaluation apprehension," 109
evolutionary theory, 42–43
expertise, and creativity, 51–52, 104

failure. See also risk
    acceptable failure and, 163–164
    admission of risk of, 179–181
    case of, 120–124
    fear of, 163–164
    preparation for, 124–126, 127–141
    resource preservation and, 131–135
    reward system and, 128–131
    staying motivated and, 135–141
    succeeding in the face of, 127–141
    success of innovators and, 95–96
fear, 163–164
    acknowledgment of, 179–181
    physiological responses to, 180
    sources of, 173–179
field
    vs. intersection of fields, 16–17
    success in, and inertia, 173–176
    as term, 16
    time spent in, and inertia, 176–177
Fields, Randolph, 161
Fiorina, Carly, 172
"flash-in-the-sky serendipity," 70–71
France Telecom, 143–144, 148
"free rider phenomenon," 109
Frensh, Peter, 51–52
Friedman, Tom, 21–22
fun, 178

Gadiesh, Orit, 75–77, 105
Garfield, Richard, 61–66, 67, 71, 72, 86,

113, 156, 163, 188
Gladwell, Malcolm, 18, 70
glass, myth about, 183–184
globalization, 22–24
Gorbachev, Mikhail, 155
Gould, John, 42–43
groups. See teams
Guilford, J. P., 40, 106

Hagstrom, Robert, 28
Harley-Davidson, 148
Hawkins, Jeff, 131–132
health care, 186–187
Heeger, Allen, 165
Hellman, Martin, 54–55
Herbert, Frank, 77–78
Highland Capital, 49–53
HSBC (bank), 46

Icosystem (company), 145
IDEO (design shop), 107, 109
incentive system
    effect on creativity, 136–138, 139–140
    failure and, 128–131
incubation period, 71, 114
innovation. See also creativity
    characteristics of, 13–15
    intersection as best opportunity for, 20
innovators. See also Darwin, Charles;
        Garfield, Richard; Lans, Håkan;
        Prothrow-Stith, Deborah; Samuels-
        son, Marcus
    characteristics of, 92–94
    factors in productivity of, 95–97
    planning for failure and, 124–126,
        127–141
    productivity of, 90–92, 94
    staying in the Intersection and,
        101–102
Internet, 30–32
    commerce on, 54–55
    motivation and, 138–139

Internet (*continued*)
  start-ups and, 134
Intersection
  courage and, 164–166
  daring to step into, 189–190
  defined, 16–17
  fear of risks at, 163–164
  as locus of future, 186–187
  as low-risk, 169, 179
  motivation at the, 136–138
  power of the, 97–101
Intersectional ideas, 19–20
  action and, 114–116
  brainstorming and, 105–111
  characteristics of, 19–20
  computational leaps and, 28–32
  depth vs. breadth and, 104–105
  directional ideas and, 17–20. *See also*
    directional vs. Intersectional ideas
  diverse teams and, 79–83, 104–105
  evaluation and, 111–114
  expecting the unexpected and, 187–188
  factors in rise of, 21–32
  failure and, 120–124
  intersection hunting and, 84–86
  logic and, 188–189
  occupation diversification and, 74–78
  population movement and, 22–24
  quantity of ideas and, 95–97
  randomness and, 84–86, 95, 96
  scientific convergence and, 25–28
  seeking out, 189–190
  time pressure and, 111–114
investment/philanthropic strategies, 5, 13

Jobs, Steve, 29, 53, 150. *See also* Pixar
Johnson, Robert, 69
jokes, 67–69

Kahneman, Daniel, 174, 177–178
Kamen, Dean, 128
Kaufman, Michael, 5
Kelley, Tom, 107, 109

Keynes, John Maynard, 164
Kimera (Brooklyn store), 24
King, Stephen, 138
Koestler, Arthur, 67, 69
Konarka (company), 164–165
Kuhn, Thomas, 52

Lans, Håkan, 89–90, 92–94, 113, 115–116,
    140–141, 156–157
learning
  associative barriers and, 49–53, 92
  from mistakes, 129, 131
  perspective on risk and, 178
Leonardo da Vinci, 21, 58
Leshner, Alan, 26–27
Linux operating system, 157–159

MacGyver (TV show), 111–112
Maeder, Paul, 49–53
Magic: The Gathering (card game),
    61–66, 71, 72, 98, 113, 115, 157
Maier, N. R., 66–67
Medici Effect, as term, 2–3. *See also*
    Intersection; Intersectional ideas
Mednick, Sarnoff, 69
memes, 17–18
Michalko, Michael, 55–56, 84–85
Miller, Steve, 79–80
mind-body medicine, 154–156
mindreading experiment, 11–13, 21, 25
mistakes. *See* failure
motivation, 135–141
multidisciplinary approach
  in policymaking, 186–187
  in science, 25–28, 50–51
musicians, and value networks, 149,
    151

network paradox, 146–147
networks
  decreasing dependence on, 154–157
  need to break away from, 149–152

network paradox and, 146–147
reasons for building, 147–149
Nexia, 25
notebooks, 113–114, 132
novelty, and creative ideas, 14

Oates, Joyce Carol, 91
occupation diversification, 74–78
Oldfield, Mike, 98–102, 105, 107
"Open Societies," 5
optical fibers, 184
Orowan, Egon, 183
Osborn, Alex, 108

Palm Pilot, 131–132
passion, 185
Pasteur, Louis, 71–72
Pauling, Linus, 26, 103
Pearce, Mick, 3–4, 13
perspectives
    diversity of, 57–60
    risk and, 177–179
Peter's Café, 1–2
Picasso, Pablo, 91, 132
Pixar, 28–29, 149–151
Poe, Edgar Allan, 84
policymaking, 186–187
pollution rights, trade in, 58–59
population movement, 22–24
"prepared-mind discoveries," 71–72
Prince (entertainer), 91
productivity
    factors in, 95–97
    of innovators, 90–92, 94
    quality and, 90–92, 96–97
prospect theory, 174
Prothrow-Stith, Deborah, 119–124, 135,
    141, 154, 157, 188

quantity of ideas
    quality and, 90–92, 96–97
    rewarding failure and, 128–131

randomness
    creativity and, 70–72, 95
    intersection hunting and, 84–86
realization, and innovation, 15
recognition, 140–141
Renaissance Florence, 2–3, 20, 21
resources
    enough, 168–169
    preservation of, 131–135
rewards. See incentive system
risk
    balanced view of, 171–182
    behavioral traps and, 172–179
    Intersectional vs. directional ideas
        and, 163–164, 177–179
    minimization of, 164–168
risk homeostasis, 167–169
Rivest, Ronald, 55
RSA cipher, 55

Samuelsson, Marcus, 35–38, 40–41,
    47–49, 60, 102, 105, 114, 156, 178
Santa Fe Institute (SFI), 27–28, 144–145
Sato, Vicki, 130
scientific discovery, nature of, 25–28
self-education, 52–53
selfish gene, 17–18
Sequoyah, 22
serial entrepreneurship, 164–166
Serruya, Mijail, 11–12, 13, 105
SFI. See Santa Fe Institute
Shakira (musician), 23–24
Shamir, Adi, 55
similar-attraction effect, 80–82
Simonton, Dean Keith, 6, 96–98, 195n3
Skinner effect, 136
small-worlds phenomenon, 28
Soros, George, 5, 13
spider silk, 25
Staley, Warren, 31
STDMA (Self-organizing Time Division
    Multiple Access), 92–93
Sternberg, Robert, 51–52
Stroebe, Wolfgang, 109

sunk costs, 176–177
Susskind, Larry, 181–182
Sutton, Robert, 6, 82, 129
Swahn, Håkan, 35, 36, 47–49, 60
"swarm intelligence," 145

Tanenbaum, Andrew, 158–159
teams
    depth vs. breadth of ideas and,
        104–105
    diversity and, 79–83
    enhancement of brainstorming by,
        110–111
    problems with group brainstorming
        and, 108–110
    time pressure and, 111–114
Theraulaz, Guy, 143–144, 188
thought walk, 85–86
time
    for evaluation, 111–114
    for failure, 135
    sunk costs and, 176–177
TMDA (Time Division Multiple Access), 90
Torvalds, Linus, 157–159
Tracy, Mark, 30–32
truck routing, 145–146
Tubular Bells (album), 98–102, 107

Turner, Jeffrey, 25
Tversky, Amos, 174, 177–178

UAVs. *See* unmanned aerial vehicles
uncertainty, 133–135, 163. *See also* failure
unmanned aerial vehicles (UAVs), 145

vaccination, 71–72
value, and creative ideas, 14–15
value networks, 147–149
    intersectional innovation and, 149–152
Vertex Pharmaceuticals, 129–130
violence-prevention strategies, 120–124
Virgin Airlines, 161–162
Virgin Group, 98–102

Walt Disney Company, 150, 151, 193n16
Wilde, Gerald, 167
Wizards of the Coast (company), 61, 86
word associations, 40–41. *See also* associative barriers
workplace, diversity in, 82–83

Zagat, Tom, 36

About the Author

FRANS JOHANSSON is a consultant who writes and speaks about intersections of all types. He was previously a founder and CEO of Inka.net, a Boston-based enterprise software company, and vice president of business development of Dola Health Systems, a health care company operating in Baltimore and Sweden. He has written on a diverse range of topics, from business management to fishing to fantasy-gaming adventures.

Johansson earned his M.B.A. from Harvard Business School and his S.B. in environmental science from Brown University. Born and raised in Sweden, he currently resides in New York City.

Create <u>Intersection</u>
between Functions, BU's
to <u>create</u> new ideas.

- Supply Chain + Mktg.
  (consumer driven opers)
- HR + Finance / Planning
              Economic driven
  (Succession Planning)

- US + Canada
  - Optimize common
    title performance